Tara Auxt Baugher
Suman Singha
Editors

Concise Encyclopedia
of Temperate Tree Fruit

More pre-publication
REVIEWS, COMMENTARIES, EVALUATIONS . . .

"*Concise Encyclopedia of Temperate Tree Fruit* contains a wealth of information on all aspects of temperate tree fruit science. Unlike many tree fruit publications, it is easy to read and well organized. It will be an excellent source of sound science-based information for serious students, commercial growers, scientists, and allied industry personnel. A complete list of literature citations is contained at the end of each chapter for those who wish to read more. This book will also be a good source of information for the serious gardener and backyard fruit grower. If we follow good maturity standards and harvest techniques described in the *Concise Encyclopedia of Temperate Tree Fruit*, we will all enjoy the very best in quality."

Jerome L. Frecon, MS
Agricultural Agent/Professor,
Rutgers University

"This book represents a very broad coverage of many topics related to temperate tree fruit culture and production. While the bulk of the information presented deals with apple, other temperate tree fruits (pear, peach, sweet cherry, and other stone fruits) are frequently mentioned throughout the text. The book is organized alphabetically into 42 chapters, each prepared by an expert in the particular subject matter. Each chapter is well written and presents its information in a clear and easy-to-read format, along with additional illustrative figures and tables. The range of topics is truly encyclopedic, and includes fruit tree structure, physiology, growth and development, soil and weather factors, pests and diseases, and nutritional value of fruits. Readers who will benefit most from this book should have had a basic introduction to chemistry, physics, and plant science to appreciate the terminology used throughout the book. For those interested in more information on any topic, each chapter is followed by suggested references for more in-depth study. This book will be a valuable source of information for students, teachers, gardeners, growers, tree fruit industry people, and anyone else with an interest in any aspect of tree fruit production."

Don C. Elfving, PhD
Horticulturist and Professor,
Washington State University,
Tree Fruit Research & Extension Center,
Wenatchee, WA

Concise Encyclopedia of Temperate Tree Fruit

FOOD PRODUCTS PRESS®
Crop Science
Amarjit S. Basra, PhD
Senior Editor

Mineral Nutrition of Crops: Fundamental Mechanisms and Implications by Zdenko Rengel

Conservation Tillage in U.S. Agriculture: Environmental, Economic, and Policy Issues by Noel D. Uri

Cotton Fibers: Developmental Biology, Quality Improvement, and Textile Processing edited by Amarjit S. Basra

Heterosis and Hybrid Seed Production in Agronomic Crops edited by Amarjit S. Basra

Intensive Cropping: Efficient Use of Water, Nutrients, and Tillage by S. S. Prihar, P. R. Gajri, D. K. Benbi, and V. K. Arora

Physiological Bases for Maize Improvement edited by María E. Otegui and Gustavo A. Slafer

Plant Growth Regulators in Agriculture and Horticulture: Their Role and Commercial Uses edited by Amarjit S. Basra

Crop Responses and Adaptations to Temperature Stress edited by Amarjit S. Basra

Plant Viruses As Molecular Pathogens by Jawaid A. Khan and Jeanne Dijkstra

In Vitro Plant Breeding by Acram Taji, Prakash P. Kumar, and Prakash Lakshmanan

Crop Improvement: Challenges in the Twenty-First Century edited by Manjit S. Kang

Barley Science: Recent Advances from Molecular Biology to Agronomy of Yield and Quality edited by Gustavo A. Slafer, José Luis Molina-Cano, Roxana Savin, José Luis Araus, and Ignacio Romagosa

Tillage for Sustainable Cropping by P. R. Gajri, V. K. Arora, and S. S. Prihar

Bacterial Disease Resistance in Plants: Molecular Biology and Biotechnological Applications by P. Vidhyasekaran

Handbook of Formulas and Software for Plant Geneticists and Breeders edited by Manjit S. Kang

Postharvest Oxidative Stress in Horticultural Crops edited by D. M. Hodges

Encyclopedic Dictionary of Plant Breeding and Related Subjects by Rolf H. G. Schlegel

Handbook of Processes and Modeling in the Soil-Plant System edited by D. K. Benbi and R. Nieder

The Lowland Maya Area: Three Millennia at the Human-Wildland Interface edited by A. Gómez-Pompa, M. F. Allen, S. Fedick, and J. J. Jiménez-Osornio

Biodiversity and Pest Management in Agroecosystems, Second Edition by Miguel A. Altieri and Clara I. Nicholls

Plant-Derived Antimycotics: Current Trends and Future Prospects edited by Mahendra Rai and Donatella Mares

Concise Encyclopedia of Temperate Tree Fruit edited by Tara Auxt Baugher and Suman Singha

Concise Encyclopedia of Temperate Tree Fruit

Tara Auxt Baugher
Suman Singha
Editors

CRC Press
Taylor & Francis Group
Boca Raton   London   New York

CRC Press is an imprint of the
Taylor & Francis Group, an informa business

Reprinted 2010 by CRC Press

CRC Press
6000 Broken Sound Parkway, NW
Suite 300, Boca Raton, FL 33487
270 Madison Avenue
New York, NY 10016
2 Park Square, Milton Park
Abingdon, Oxon OX14 4RN, UK

Published by

Food Products Press® and The Haworth Reference Press, imprints of The Haworth Press, Inc., 10 Alice Street, Binghamton, NY 13904-1580.

TR: 3.9.04

DISCLAIMER
Information on private enterprises or products does not imply endorsement by the writers or by The Haworth Press, Inc.

Cover design by Lora Wiggins.

Library of Congress Cataloging-in-Publication Data

Concise encyclopedia of temperate tree fruit / Tara Auxt Baugher, Suman Singha, editors.
      p. cm.
   Includes bibliographical references and index.
   ISBN 1-56022-940-3 (alk. paper) — ISBN 1-56022-941-1 (alk. paper)
   1. Fruit-culture—Encyclopedias. 2. Fruit trees—Encyclopedias. I. Baugher, Tara Auxt. II. Singha, Suman.

SB354.4 .C66 2003
634'.03—dc21

                                                                                      2002028814

To Allison, Javed, Kamini, Kathy, and Phil

CONTENTS

List of Figures

List of Tables

ABOUT THE EDITORS

Tara Auxt Baugher, PhD, received her initial training in pomology from her mother and grandfather, who grew apples and peaches in Paw Paw, West Virginia. She has a BA from Western Maryland College and MS and PhD degrees from West Virginia University. From 1980 to 1994, she was a West Virginia University tree fruit extension specialist and horticulture professor, and she is currently a tree fruit consultant in south-central Pennsylvania. Dr. Baugher conducts research in the areas of intensive orchard management systems and sustainable fruit production and has published more than 30 papers in refereed journals and books. Dr. Baugher's honors and awards include the Gamma Sigma Delta Award of Merit, West Virginia University's Outstanding Plant Science Teacher, and the International Dwarf Fruit Tree Association Researcher of the Year.

Suman Singha, PhD, was raised on an apple orchard in northern India. He received BS and MS degrees from Punjab Agricultural University and a PhD from Cornell University. He served on the faculty of West Virginia University from 1977 to 1990, and was promoted to Professor of Horticulture in 1986. In 1990, he accepted the position of Professor and Head of the Department of Plant Science at the University of Connecticut, and in 1995 was appointed Associate Dean of the College of Agriculture and Natural Resources. Dr. Singha was named an American Council on Education Fellow in 2000. His research interests focus on tissue culture and fruit tree physiology and management. Dr. Singha has been recognized for teaching excellence and for contributions to first-year student programs.

CONTRIBUTORS

Rajeev Arora, PhD, Environmental Stress Physiologist, Associate Professor of Horticulture, Iowa State University, Ames, Iowa.

Andrea T. Borchers, PhD, Nutritionist and Scientific Writer, Sonoma, California.

L. L. Creasy, PhD, Pomologist, Professor Emeritus, Cornell University, Ithaca, New York.

Mervyn C. D'Souza, PhD, Food Technologist, Director of Technical Services, Knouse Foods Cooperative, Inc., Biglerville, Pennsylvania.

David C. Ferree, PhD, Pomologist, Professor of Horticulture and Crop Science, Ohio State University, Ohio Agricultural Research and Development Center, Wooster, Ohio.

James A. Flore, PhD, Tree Fruit Physiologist, Professor of Horticulture, Michigan State University, East Lansing, Michigan.

D. Michael Glenn, PhD, Soil Scientist, Research Soil Scientist, USDA-ARS Appalachian Fruit Research Station, Kearneysville, West Virginia.

Martin C. Goffinet, PhD, Developmental Anatomist, Senior Research Associate, Cornell University, New York State Agricultural Experiment Station, Geneva, New York.

Duane W. Greene, PhD, Pomologist, Professor of Horticulture, University of Massachusetts, Amherst, Massachusetts.

John M. Halbrendt, PhD, Nematologist, Assistant Professor of Nematology, Pennsylvania State University Fruit Research and Extension Center, Biglerville, Pennsylvania.

Peter M. Hirst, PhD, Pomologist, Associate Professor of Pomology, Purdue University, West Lafayette, Indiana.

Dianne A. Hyson, PhD, RD, Nutritionist, Assistant Professor of Family and Consumer Sciences, California State University, Sacramento, California.

Alan N. Lakso, PhD, Fruit Crop Physiologist, Professor of Horticulture, Cornell University, New York State Agricultural Experiment Station, Geneva, New York.

Tracy C. Leskey, PhD, Behavioral Entomologist, Research Entomologist, USDA-ARS Appalachian Fruit Research Station, Kearneysville, West Virginia.

Ian A. Merwin, PhD, Pomologist and Agroecologist, Associate Professor of Horticulture, Cornell University, Ithaca, New York.

Stephen S. Miller, PhD, Pomologist, Research Horticulturist, USDA-ARS Appalachian Fruit Research Station, Kearneysville, West Virginia.

Stephen C. Myers, PhD, Pomologist, Professor and Chair of Horticulture and Crop Science, Ohio State University, Columbus, Ohio.

Desmond O'Rourke, PhD, Agricultural Economist, Emeritus Professor of Agricultural Economics, Washington State University, and President, Belrose, Inc., Pullman, Washington.

Katharine B. Perry, PhD, Agricultural Meteorologist, Professor of Horticultural Science and Assistant Dean, College of Agriculture and Life Sciences, North Carolina State University, Raleigh, North Carolina.

A. Nathan Reed, PhD, Postharvest Physiologist, Assistant Professor of Horticulture, Pennsylvania State University Fruit Research and Extension Center, Biglerville, Pennsylvania.

Curt R. Rom, PhD, Pomologist, Associate Professor of Horticulture, University of Arkansas, Fayetteville, Arkansas.

David A. Rosenberger, PhD, Plant Pathologist, Professor of Plant Pathology, Cornell University, Hudson Valley Laboratory, Highland, New York.

Ralph Scorza, PhD, Tree Fruit Breeder and Geneticist, Research Horticulturist, USDA-ARS Appalachian Fruit Research Station, Kearneysville, West Virginia.

Dariusz Swietlik, PhD, Tree Fruit Physiologist, Director, USDA-ARS Appalachian Fruit Research Station, Kearneysville, West Virginia.

Christopher S. Walsh, PhD, Pomologist, Professor of Horticulture, University of Maryland, College Park, Maryland.

Christopher B. Watkins, PhD, Postharvest Physiologist, Professor of Horticulture, Cornell University, Ithaca, New York.

Preface

Tree fruit production and the associated areas of science and technology have undergone momentous transformations in recent years. Growers are adopting new cultivars, planting systems, integrated management programs, and fruit storage and marketing practices. The changes have resulted in an intensified need to increase basic and applied knowledge of fruit physiology and culture. A commitment to lifelong learning is essential for those who want to succeed in this field.

The *Concise Encyclopedia of Temperate Tree Fruit* is addressed to individuals who aspire to learn more about both the science and art of tree fruit culture. All aspects of pomology are covered, ranging from the critically important but often overlooked topic of site selection and preparation to the role of biotechnology in breeding programs. We recognize that it is difficult in a book of this breadth to adequately discuss minor crops, and thus the emphasis is on the primary tree fruit crops of the temperate zone.

To facilitate use, topics are listed alphabetically and covered in sufficient detail to provide the reader with the most significant and current information available. Related topics and selected references are provided at the end of each section for those who desire to explore a subject in greater depth. As with any concise reference book, the objective has been to make the subject matter comprehensive yet succinct.

We thank the group of outstanding contributors who made this project possible. Each is recognized as an authority in a particular research area and enthusiastically contributed his or her knowledge to making this a fine encyclopedia. We also extend appreciation to Susan Schadt and John Armstrong for their assistance with illustrations, Martin Goffinet for reviewing the text on anatomy and taxonomy, and the many educators and industry professionals who provided figures or information for tables. Finally, we acknowledge our grandfathers

for nurturing our interests in horticulture. They were dedicated orchardists who served as teachers and mentors during our formative
years. It was this common background that fostered the beginning of
a valued professional relationship between the two of us.

One of the greatest rewards of a vocation in pomology is working
with individuals who are genuinely committed to finding novel ways
to modernize agriculture, improve human nutrition, and safeguard
farmlands. We offer this book as a tribute to the students, growers,
and scientists whose collaborative efforts lead to advancements in
feeding and sustaining our world.

1. ANATOMY AND TAXONOMY

Anatomy and Taxonomy

Tara Auxt Baugher

Nomenclature, classification, and description are the basic components of systematic pomology. The identification or study of a fruit species involves detailed examination of distinguishing anatomical characteristics, such as leaf shape, inflorescence type, and fruit type (Figure A1.1). International codes of nomenclature govern family, genus, and species taxa.

POME FRUIT

Apple (*Malus* Mill.)

Family Rosaceae, Subfamily Pomoideae; approximately 30 species; domestic apples derived mainly from *M. pumila* Mill.; domestic crab apples, hybrids of *M. pumila* and *M. baccata* (L.) Borkh. or other primitive species; hybrids numerous and complex.

Deciduous, infrequently evergreen, branching tree or shrub; leaves folded or twisted in buds, ovate or elliptic or lanceolate or oblong, lobed or serrate or serrulate; buds ovoid, a few overlapping scales.

Flowers white to pink or crimson, epigynous, in cymes; stamens 15 to 50; styles 2 to 5; ovary 3 to 5 cells.

Fruit a pome, oblong or oblate or conic or oblique, diameter 2 to 13 centimeters, various hues of green to yellow to red, varying russet and lenticel characteristics; flesh lacking stone cells.

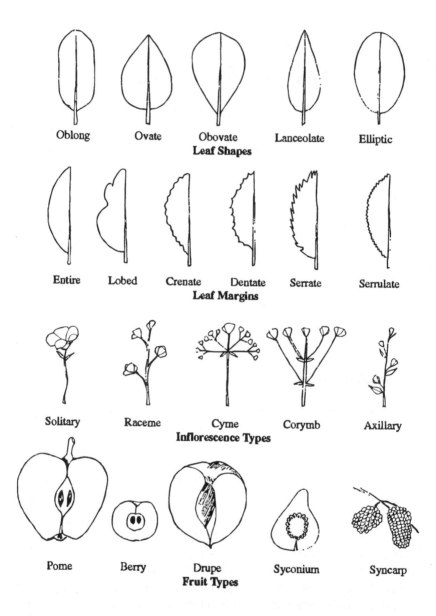

FIGURE A1.1. Illustrations of terminology used to describe temperate tree fruit (*Source:* Modified from Harris and Harris, 1997, *Plant Identification Terminology;* and Bailey et al., 1976.)

Pear (*Pyrus* L.)

Family Rosaceae, Subfamily Pomoideae; approximately 20 species; European (*P. communis* L.) possibly derived from *P. caucasica* Fed. and *P. nivalis* Jacq.; Asian mostly from *P. pyrifolia* (Burm. f.) Nakai and *P. ussuriensis* Maxim. selections; hybrids numerous and complex.

Deciduous or semievergreen tree; leaves rolled in buds, ovate or oblong or elliptic or lanceolate or obovate, crenate or serrate or entire, infrequently lobed; buds ovoid, overlapping scales.

Flowers, white, sometimes with pink tinge, epigynous, in corymbs, open with or before leaves; stamens 20 to 30; styles 2 to 5; ovules 2 per cell.

Fruit a pome, pyriform or globose or ovoid, diameter 2 to 12 centimeters, various hues of green/yellow or red/brown, varying russet and lenticel characteristics; flesh usually containing grit or stone cells.

Quince, Common (*Cydonia oblonga* Mill.)

Family Rosaceae, Subfamily Pomoideae.

Deciduous small tree or shrub; leaves ovate or oblong, entire, tomentose on underside; buds pubescent, small.

Flowers pink or white, epigynous, solitary, terminal on leafy shoots; stamens 20 or more; styles 5; ovary 5 cells.

Fruit a pome, pyriform or globose, hard, diameter 4 to 8 centimeters, yellow, many-seeded.

STONE FRUIT

Peach/Nectarine (*Prunus persica* (L.) Batsch.)

Family Rosaceae, Subfamily Prunoideae, Subgenus *Amygdalus* (L.) Focke.

Deciduous, small tree; branches glabrous; leaves alternate, folded in buds, long-lanceolate, serrulate; buds 3 per axil, the 2 laterals being flower buds.

Flowers pink to red, perigynous, sessile or on short stalk, open before leaves.

Fruit a drupe, peach tomentose, nectarine glabrous, globose or oval or oblate, sometimes compressed, diameter 4 to 10 centimeters, yellow to red; stone sculptured or pitted.

Almond (*Prunus amygdalus* Batsch.)

Family Rosaceae, Subfamily Prunoideae, Subgenus *Amygdalus* (L.) Focke.

Deciduous, spreading tree; branches glabrous; leaves alternate, folded in buds, lanceolate, serrulate; buds 3 per axil, the 2 laterals being flower buds.

Flowers white to pink, perigynous, sessile or on short stalk, open before leaves.

Fruit a drupe, tomentose, oblong, compressed, length 2 to 6 centimeters, green, dry; stone pitted; kernel sweet.

Apricot (*Prunus armeniaca* L.)

Family Rosaceae, Subfamily Prunoideae, Subgenus *Prunophora* Focke.

Deciduous, small tree; bark reddish; branches glabrous; leaves alternate, usually ovate, serrate, pubescent veins on underside.

Flowers white or pink, perigynous, solitary, open before leaves.

Fruit a drupe, pubescent when immature, almost smooth when mature, oblong or globose, sometimes compressed, diameter 2 to 6 centimeters, yellow, sometimes red cheek; stone smooth, flattened, ridged suture.

Plum, Common or European (*Prunus domestica*; also *P. cerasifera, P. spinosa,* and *P. insititia*); Japanese or Oriental (*P. salicina*); Native American (many species, including *P. americana* or wild plum)

Family Rosaceae, Subfamily Prunoideae, Subgenus *Prunophora* Focke.; numerous species and hybrids.

Deciduous, small tree; leaves alternate, usually ovate/obovate or oblong/elliptic, often serrate or crenate.

Flowers usually white, perigynous, solitary or clustered, open before or sometimes with leaves.

Fruit a drupe, glabrous, usually with bloom on skin, length 2 to 8 centimeters, globose or oblong or cordate or elliptical, sometimes compressed, variable colors, including blue/purple or red/pink or yellow/green; stone smooth, flattened.

Cherry, Sour (*Prunus cerasus* L.); Sweet (*P. avium* L.); Duke (hybrid, *P. avium* x *P. cerasus*); Native American (*P. besseyi* L. H. Bailey or sand cherry, and others)

Family Rosaceae, Subfamily Prunoideae, Subgenus *Cerasus* Pers.; various species and hybrids.

Deciduous, small to large tree; leaves alternate, folded in buds, ovate or obovate or obovate/elliptic, serrate or crenate/serrate.

Flowers white to pink, perigynous, solitary or in inflorescences ranging from few-flowered umbels to racemes.

Fruit a drupe, globose or oblate or cordate, diameter 1 to 3 centimeters, dark or light red or yellow; stone sculptured or smooth.

OTHER TEMPERATE TREE FRUIT

Fig, Common (*Ficus carica* L.)

Family Moraceae.

Deciduous, small, irregular tree; leaves thick, deeply lobed (3 to 5), pubescent on underside, scabrous on top.

Flowers small, solitary and axillary.

Fruit a syconium, pyriform, diameter 2 to 6 centimeters, greenish purple/brown, many-seeded.

Mulberry (*Morus* L.)

Family Moraceae; approximately 12 species.

Deciduous, open tree; leaves alternate, frequently lobed, crenate or serrate or dentate, scabrous or glabrous; buds of 3 to 6 overlapping scales.

Flowers unisexual, male and female in separate inflorescences, usually on separate trees (dioecious), sometimes on the same tree (monoecious), appear with leaves.

Fruit a syncarp, ovoid or cylindric, length 1 to 4 centimeters, red/purple or pink/white, resembles a blackberry.

Papaw, Northern (also Pawpaw; *Asimina triloba* (L.) Dunal.)

Family Annonaceae.
Deciduous, small tree; leaves alternate, oblong and obovate, drooping.
Flowers purple, on hairy stalks, axillary, open before leaves.
Fruit a berry, usually oblong or elliptical, length 7 to 12 centimeters, usually green/yellow to brown/bronze when ripe; seeds compressed.

Persimmon (*Diospyros* L.)

Family Ebenaceae.
Deciduous or evergreen, dioecious tree or shrub; leaves alternate, variable shapes; buds ovoid, a few outer scales.
Flowers usually white or yellow, unisexual, female solitary, male in cymes.
Fruit a berry, usually globose or oblate, diameter 2 to 10 centimeters, usually orange/yellow, turning brown/black, large persistent calyx; seeds large, compressed, 1 to 10.

Jujube, loquat, medlar, pomegranate, and serviceberry also are grown in temperate zones. These shrubs or small trees are described in an encyclopedia on small fruit, to be edited by Robert Gough and published by The Haworth Press, Inc.

Knowledge of taxonomy and anatomy is important in studies on fruit physiology and culture. Moreover, classification and description systems have practical applications, such as ensuring graft compatibility and increasing fruit set.

Related Topics: CULTIVAR SELECTION; FRUIT GROWTH PATTERNS; ROOTSTOCK SELECTION

SELECTED BIBLIOGRAPHY

Bailey, Liberty Hyde, Ethel Zoe Bailey, and staff of the Liberty Hyde Bailey Hortorium (1976). *Hortus third: A concise dictionary of plants cultivated in the United States and Canada.* New York: Macmillan Publishing Co.
Westwood, Melvin N. (1993). *Temperate-zone pomology: Physiology and culture.* Portland, OR: Timber Press.

B

1. BREEDING AND MOLECULAR GENETICS

Breeding and Molecular Genetics

Ralph Scorza

Tree fruit have been the subjects of genetic improvement for thousands of years. From simply planting seed of the most desirable fruit, to vegetatively propagating the best trees by grafting, to cross-pollination, to the use of gene transfer and genetic mapping, humans have continuously striven for the "perfect" fruit. We have come a long way in our expectations of fruit quality and availability. From times of limited availability, and marketing and consumption of locally produced fruit, we are now in an age where high-quality, blemish-free fruit are expected year-round by most consumers in industrialized countries. Fruit may be grown thousands of miles from the point of consumption, travel weeks to market, and be stored for nearly a year. Breeders strive to develop cultivars that meet the stringent demands of growers, packers, shippers, wholesalers, retailers, and consumers.

Modern fruit-breeding objectives can be divided into two broad classes, those specifically aimed at improving the fruit, and those aimed at improving the tree. Fruit traits include size, flavor, texture, color, disease resistance, and the ability to maintain quality during and following storage. Tree traits include precocity, vigor, size, and resistance to diseases, insects, cold, heat, drought, and flooding. Although tree traits are vitally important for production efficiency, their improvement can be meaningful only if fruit quality is equal or superior to existing commercial cultivars. Therefore, tree fruit breeding has always been concerned with the selection and optimization of complex, multiple traits.

Tree fruit present both unique difficulties and unique opportunities in terms of genetic improvement when compared with herbaceous crops. Long generation cycles and high levels of genetic heterozygosity make the development of improved fruit cultivars a time-consuming process. For many fruit crops, new cultivars, especially those carrying novel traits, cannot be developed within the span of a breeder's career. For example, peach, a crop with a relatively short generation time of three to four years, has generally required from 10 to 20 years from first fruiting to cultivar release.

The relatively large land areas necessary to grow segregating populations of tree crops add considerable expense to breeding programs. High costs can limit the number of seedlings that are grown, reducing the probabilities of encountering the rare combination of genes necessary to produce a superior cultivar. Once a new cultivar is released to the market it must compete with existing cultivars. The production of some of the most economically important tree fruit relies on the use of a few cultivars. For example, 'Bartlett' accounts for approximately 50 percent of the commercial pear production in North America. Together, 'Bartlett', 'Beurre Bosc', and 'Anjou' account for almost all of the commercial production in the United States. Over 50 percent of the world apple crop is based on 'Delicious', 'Golden Delicious', 'Granny Smith', 'Gala', and 'Fuji'. Sour cherry production in the United States is based almost entirely on 'Montmorency'. These major fruit cultivars are broadly adapted, and a significant body of information exists concerning their production, storage, and marketing. New cultivars, regardless of their apparent superior qualities, lack both an information base and a record of consumer acceptance, making their introduction and adaptation by the industry difficult and slow.

Notwithstanding the difficulties in developing new, successful fruit cultivars, certain characteristics of these species aid cultivar development. Once a desirable phenotype is selected by the breeder, it can be reproduced indefinitely through vegetative propagation. No further breeding is necessary to fix traits in a population as would generally be necessary for seed-propagated herbaceous crops. Vegetative propagation may be through the rooting of cuttings, but more often it is through graftage of a scion cultivar onto a rootstock. This separation of a plant into rootstock and scion genotypes provides a flexibility whereby the scion is not selected for root characteristics (adaptation to specific soils, resistance to soilborne diseases, insects, nematodes, etc.), nor is the rootstock selected for fruiting characteristics. In this way, the selection criteria for any one cultivar of rootstock or scion can be simplified with rootstocks and scions "mixed and matched" to optimize production and quality over a range of environmental conditions.

Although characteristics under selection for rootstock or scion breeding may differ, the breeding approaches employed are the same. The general approaches to fruit breeding currently in use are hybridization, gene transfer, and molecular marker-assisted selection.

HYBRIDIZATION

Hybridization is the traditional method of tree fruit breeding and remains the dominant technology for cultivar development. This method has passed the test of time, and most of our major fruit cultivars are products of uncontrolled or controlled hybridization, with mutations of these hybrids producing additional cultivars. Uncontrolled hybridization involves little more than collecting seed from trees that may be self- or cross-pollinated or a mixture of both. Controlled hybridization is the technique of applying pollen of the selected male parent to the receptive stigma of the selected female parent. In the case of self-compatible species, such as peach, covering trees with an insect-proof material ensures self-pollination. Although simple in concept, controlled pollination requires in-depth knowledge of crop biology and relies on exacting techniques for maximizing the number of hybrid progeny. Many of these techniques are described in Janick and Moore (1975, 1996) and Moore and Janick (1983).

Seed resulting from hybridization are collected and planted. The resulting progeny is generally a heterozygous, heterogeneous population expressing diverse genetic traits. Seedlings are evaluated for the traits of interest. The desired traits are expected to result from the combination of desirable traits inherited from both parents, although undesirable traits are also inherited by the progeny. Under ideal conditions, the inheritance of the trait(s) of interest is known and there is an expected proportion of the population that carries one or more of the traits. More often, since many tree fruit traits follow a complex, multigenic pattern of inheritance, the inheritance of many traits is not known and there is little to do but evaluate the phenotype of the progeny. Such testing may require artificial inoculations to evaluate disease and insect resistance in the greenhouse, field, and in storage; evaluation of fruit, including quality and storage characteristics; and productivity data. Data are generally collected in multiple years in multiple locations before a cultivar is released for commercial use.

NONTRADITIONAL BREEDING TECHNIQUES

New biotechnologies under development and currently used for the genetic improvement of plant species have the potential to revolutionize tree fruit breeding and cultivar development. Development of new tree fruit cultivars has been plagued by generation cycles of three to 20 years, high levels of heterozygosity, severe inbreeding depression, complex intraspecific incompatibility relationships, and nucellar embryony. The application of new technologies that will speed the pace of tree improvement is critically needed.

Genetic Transformation and the Production of Transgenic Plants

The importance of genetic transformation lies in the fact that this technique allows for the insertion of a single or a few genes into an established genotype, avoiding the random assortment of genes produced through meiosis. The mixing of genetic material of both parents that results from hybridization provides genetic diversity, but it also generally requires generations of continued hybridization and selection to produce progeny carrying the desired traits of both parents. Transformation has the potential to produce a genotype that is essentially unchanged except for the improvement of a particular trait. Genetic transformation is not without its own inherent difficulties. The regeneration of many tree fruit species and/or cultivars is problematic and one of the most serious hindrances to the application of gene transfer technologies to perennial tree fruit crops. In those species that can be reliably transformed, the technology is generally only successful with a few genotypes and, in some cases, these genotypes are not commercially important. In some species, transformation has been obtained only from seedling material. This reduces the usefulness of the technology, since hybridization using transgenic genotypes as parents would then generally be required for cultivar development in order to combine the trait improved through transformation with the host of additional traits necessary for a commercial cultivar. It is precisely the process of hybridization that genetic transformation seeks to bypass. The difficulties of developing transgenic tree fruit are clearly indicated by the fact that of the 8,906 field releases of transgenic plants in the United States from 1987 to July 2002, only 54 were temperate tree fruit species.

The process of producing transgenic plants of tree fruit species can vary from months to years. Following the initial confirmation of transformation, the processes of rooting and propagation add significant time to the process. Once confirmed transgenic lines are obtained, the requirements for evaluating their performance are the same as those for conventional cultivar development. Transgenic plants require careful and exhaustive testing not only to evaluate the effect of the transgene but also to confirm the trueness-to-type and stability of the characteristics of the original cultivar.

Difficulties notwithstanding, there are several inherent advantages in the use of gene transfer for tree fruit improvement. Once a useful transformant is isolated, assuming stability of transgene expression (and this assumption has yet to be adequately tested for tree fruit), vegetative propagation—the normal route of multiplying tree fruit—provides for virtually unlimited production of the desired transgenic line. Fixation through the sexual cycle is not required. Also, while the dominance of a few major cultivars in many tree fruit crops such as pear, apple, and sour cherry may be a hindrance to the acceptance of new cultivars produced through hybridization, it can maximize the impact of a major cultivar improved through transformation.

Genetic Marker-Assisted Selection and Gene Identification

Genomics research is aimed at identifying functional genes and understanding the action and interactions of gene products in controlling plant growth and development. For plant breeders, perhaps the two most important aspects of genomics research are the development of genetic markers and gene identification (and isolation).

Molecular markers are segments of deoxyribonucleic acid (DNA) that reveal unique sequence patterns diagnosed by one or more restriction enzymes in a given cultivar, line, or species. These DNA segments are associated with genes that control plant characteristics. If the marker is found in DNA isolated from a particular plant line or cultivar, the character with which it is associated will likely be expressed in that plant. The closer the marker DNA segment is to the segment of DNA, or gene, controlling the character of interest, the stronger the association between the marker and the character. If the marker DNA is contained in the gene of interest, the association will be 100 percent. To develop markers, trees are rated for the expression of characteristics such as resistance to a particular disease, fruit size, sugar content, etc. This is a critical step in marker development. The value of a marker in predicting a trait or phenotype is a function of the accuracy of the evaluation of the trait in the mapping population. A marker cannot be accurately correlated with a characteristic if the characteristic cannot be accurately rated or quantified. DNA is extracted from the evaluated plants, and then any one of a number of molecular marker systems is used to identify polymorphisms that are preferentially present in those trees that carry the trait(s) under consideration. For a discussion of the various marker systems, see Abbott in Khachatourians et al. (2002). Once the presence of a marker or markers is associated with a trait, and particularly if this association is consistent across populations with varying genetic backgrounds, the marker can be used to predict the presence of the trait in hybrid progeny. In programs using molecular markers, the presence of marked or mapped traits can be evaluated by sampling a small amount of leaf material from young seedlings in the greenhouse, rather than waiting for seedlings to fruit or display tree characteristics in field plantings.

Molecular markers have the potential for speeding the breeding process and reducing costs. The prospect of selecting promising seedlings at an early stage of growth in the greenhouse by using a saturated genetic map to tag single gene traits as well as multigenic traits, and planting only these promising genotypes in the field, is an attractive prospect. But molecular markers are not a panacea for fruit breeding. Considerable work is involved in the development of molecular markers, particularly the initial trait evaluation in mapping populations. Also, the fact remains that hybridization will produce random gene assortment, and it is likely that several generations or more will be necessary for cultivar development (especially if genes are introduced from unimproved germplasm). Using molecular markers, seedlings with little likelihood of value can be rapidly discarded, leaving field space for more promising material. The seedlings that remain, however, must be subjected to the same rather lengthy evaluation process as previously described. For a more detailed discussion of marker-assisted selection of tree fruit, see Luby and Shaw (2001).

Gene identification utilizes the close linkage between a molecular marker and a functional gene to allow for "chromosome walking" and, ultimately, the isolation of that gene. The marker that is linked to a gene of interest can serve as an anchorage for the initiation of map-based positional cloning, from which a physical map can be established. Narrowing a gene of interest down to a limited region of a chromosome requires further fine mapping analysis to ensure the gene is located between two known molecular markers. The genomic region identified as carrying the gene is then subjected to genome sequencing, verification of complementary DNAs, identification of sequence change or rearrangement, and functional analysis (e.g., functional complementation or reverse genetics such as transfer DNA insertion or gene silencing). These analyses ensure that the gene isolated dictates the observed phenotype. The map-based approach has been successfully used for gene isolation in herbaceous species including *Arabidopsis thaliana,* maize, and tomato.

Microarray technology is used to isolate genes based on the analysis of gene expression in a particular cell type of an organism, at a particular time, under particular conditions. For instance, microarrays allow comparison of gene expression between healthy and diseased cells. Microarrays are based on the binding of complementary single-stranded nucleic acid sequences. Microarrays typically employ glass slides, onto which DNA molecules are attached at fixed locations. Ribonucleic acid (RNA) species isolated from different sources or treatments are labeled by attaching specific fluorescent dyes that are visible under a microscope. The labeled RNAs are hybridized to a microscope slide where DNA molecules representing many genes have been placed in discrete spots. By analyzing the scanned images, the change of expression patterns of a gene or group of genes, in response to different treatments or developmental stages, can be identified. The identified genes serve as potential candidates for further functional characterization.

The value of gene identification, whether through the use of molecular markers and chromosome walking, microarrays, or other technologies, is the availability of these plant-based genes for genetic transformation and targeted plant improvement. Currently, most genes available for plant transformation have been isolated from microorganisms, and only a few from plant species. The isolation and use of plant genes, particularly from woody species, for transformation will be an important step in the genetic improvement of temperate tree fruit.

The current revolution in genetics will have a dramatic effect on plant breeding. The new genetic technologies are nowhere more needed than in temperate tree fruit breeding, a process that is relatively slow and inefficient. At the same time, the relative lack of genetic information and the difficulties of regenerating most temperate tree fruit make application of the new biotechnologies most difficult for these species. The close collaboration of molecular biologists and fruit breeders will be critical to progress both in basic research on fruit genetics and the application of new knowledge and technologies to the development of new improved temperate tree fruit cultivars.

Related Topics: CULTIVAR SELECTION; PROPAGATION; ROOTSTOCK SELECTION

SELECTED BIBLIOGRAPHY

Janick, J. and J. N. Moore, eds. (1975). *Advances in fruit breeding.* West Lafayette, IN: Purdue Univ. Press.

Janick, J. and J. N. Moore, eds. (1996). *Fruit Breeding.* New York: John Wiley and Sons, Inc.

Khachatourians, G., A. McHughen, R. Scorza, W.-T. Nip, Y. H. Hui, eds. (2002). *Transgenic Plants and Crops.* New York: Marcel Dekker, Inc.

Luby, J. J. and D. Shaw (2001). Does marker-assisted selection make dollars and sense in a fruit breeding program? *HortScience* 36:872-879.

Moore, J. N. and J. R. Ballington Jr., eds. (1990). *Genetic resources of temperate fruit and nut crops,* Acta Horticulturae 290. Wageningen, the Netherlands: Internat. Soc. Hort. Sci.

Moore, J. N. and J. Janick, eds. (1983). *Methods in fruit breeding.* West Lafayette, IN: Purdue Univ. Press.

Oliveira, M. M., C. M. Miguel, and M. H. Raquel (1996). Transformation studies in woody fruit species. *Plant Tissue Cult. Biotech.* 2:76-91.

Schuerman, P. L. and A. Dandekar (1993). Transformation of temperate woody crops: Progress and potentials. *Scientia Hort.* 55:101-124.

Scorza, R. (1991). Gene transfer for the genetic improvement of perennial fruit and nut crops. *HortScience* 26(8):1033-1035.

Scorza, R. (2001). Progress in tree fruit improvement through molecular genetics. *HortScience* 36(5):855-858.

Singh, Z. and S. Sansavini (1998). Genetic transformation and fruit crop improvement. *Plant Breeding Rev.* 16:87-134.

1. CARBOHYDRATE PARTITIONING AND PLANT GROWTH
2. CULTIVAR SELECTION

Carbohydrate Partitioning and Plant Growth

Alan N. Lakso
James A. Flore

Plants convert sunshine radiant energy into chemical energy by the process of photosynthesis, and the first stable compounds formed are carbohydrates. As sugars, carbohydrates are transported within a plant and used both as building blocks of structure and for energy for respiration to fuel growth. Some sugars are synthesized into cellulose and related compounds that make up the great majority of the structure of plants. Energy is also stored in the form of sugars that can be used quickly by respiration or in polysaccharides such as starch that provide reserve stores of energy. For these reasons, carbohydrates are the most critical and ubiquitous compounds in plants. Carbohydrate partitioning (i.e., distribution) determines amounts and patterns of plant growth and yields of crops. Consequently, much research has been done to document and understand these patterns and regulation of the production and partitioning of carbohydrates in plants.

SEASONAL PATTERNS OF CARBOHYDRATE PRODUCTION

Before considering the partitioning of carbohydrates, seasonal production of carbohydrates must be understood. Temperate tree fruit are perennial plants with permanent structures that provide physical frameworks of canopies and root systems that do not need to be reproduced each season, as in annuals. Additionally, these structures contain reserves of carbohydrates and mineral nutrients that can immediately be used for early growth. Temperate tree fruit can develop full canopies and intercept sunlight to produce carbohydrates very early in the season compared to annual crops because they have thousands of preformed buds. Indeed, orchards may almost reach full light interception at the time many annual crops are planted. This leads to a rapid increase in total carbohydrate production early in the season.

In midseason, once tree canopies are established, carbohydrate production tends to be relatively stable. Compared to annuals that are planted at very close spacings and develop very high canopy densities, temperate tree fruit tend to have lower midseason carbohydrate production rates. This is due to the necessary alleyways that lower maximum sunlight interception to values typically about 40 to 70 percent for most mature orchards. Late in the season, however, temperate tree fruit tend to maintain carbohydrate production due to the ability of their leaves to sustain good photosynthesis rates for several months, while the leaves of annual crops typically senesce. Even with slower decline of leaf function, the shorter, cooler days combined with leaf aging gradually reduce carbohydrate production in the fall. This extended life span of green, active leaf area is referred to as a long "leaf area duration" and has been well correlated to high dry matter production in crops. Consequently, fruit trees often do well in marginal climates.

Photosynthesis produces carbohydrates for growth and energy. The energy needed to drive growth and maintenance of a tree is generated by using carbohydrates. The percentage of fixed carbohydrates used for respiration over a season will vary with the activity of growth, but it is reported to be as much as 40 to 60 percent in some annual plants. Unfortunately, we do not have good estimates of the respiration of root systems due to the difficulty of measurement, especially in field soils. Nor do we have good estimates of the amounts of carbohydrates lost because of leaching. However, a mature temperate fruit tree tends to have a relatively low respiration rate since (1) the structure of the tree is already built, (2) the woody structure has a high percentage of dead cells (wood vessels, fibers, etc.) that do not require much energy for maintenance, (3) the leaves tend to live long so the tree does not need to constantly produce energy-expensive new leaves to be productive, and (4) the fruit primarily accumulate carbohydrates directly so there are relatively few costs related to synthesizing proteins, lipids, etc. Combined, these characteristics make carbohydrate production and utilization quite efficient in temperate fruit trees.

PARTITIONING TO TREE ORGANS

To grow and produce a crop, all the critical organs of a fruit tree must receive carbohydrates for growth and maintenance. A unique feature inherent to perennial crops is that flower buds for the next year's crop are developing on a tree during the growth of the current year's crop. So although growers and consumers are interested in the fruit component, the perennial nature requires that vegetative organs (shoots, roots, and structure) and developing flower buds receive an adequate share of carbohydrates to sustain cropping in following years. There have been many studies of the actual partitioning of dry matter (not precisely carbohydrates, but a close approximation) in temperate tree fruit.

The individual organs on a fruit tree have genetically programmed growth patterns and carbohydrate requirements. However, it appears that overall partitioning of carbohydrates within a tree is not a genetically programmed process, but a result of the unique combination of competing organs and their relative abilities to vie for limited carbohydrates. The degree of competition among organs is determined by organ activity and distance from carbohydrate source.

In a young tree that is not yet cropping, shoots and roots receive substantial amounts of carbohydrates, as the development of both the top and bottom of the tree is needed. The relative amounts that roots receive, however, tend to decline with tree age. Once a tree begins to reach its final size (either natural or contained by grower pruning) and begins to produce fruit, the patterns of partitioning change (Figure C1.1).

The timing of growth and thus requirement for carbohydrates varies among tree organs. Shoot and leaf area growth is generally strongest in early season, with varying levels of decline in midseason as canopy development is completed. Shoot growth in apple generally declines markedly in midseason, while in stone fruit, shoot growth can continue quite strongly even after harvest. Shoots and the trunk continue to increase in diameter after terminal bud set until leaf fall.

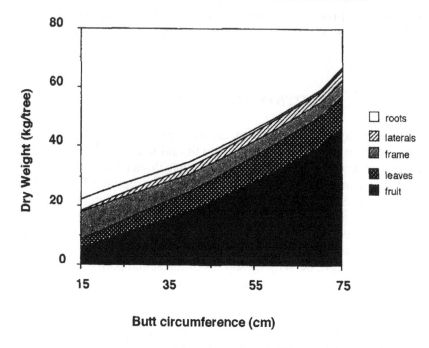

Butt circumference (cm)

FIGURE C1.1. Growth in peach tree dry weight and the partitioning into tree organs as affected by tree development over many years (*Source:* Flore and Layne, 1996; Figure 3, from original data of D. Chalmers and B. v. d. Ende, 1975, *Annals of Botany* 39:423-432. Reprinted from Eli Zamski and Arthur Schaffer, eds., *Photoassimilate Distribution in Plants and Crops: Source-Sink Relationships,* p. 830, by courtesy of Marcel Dekker, Inc.)

Patterns of fruit growth differ quite markedly between pome and stone fruit (see FRUIT GROWTH PATTERNS for details). Pome fruit growth rate (weight gain per day) increases rapidly in the first third of the season, then levels off and is fairly stable until harvest. Stone fruit growth rate increases initially similar to pome fruit, but then shows a decline in midseason, followed by another peak before harvest. The final stage of growth is very accelerated and many fruit will increase 40 to 60 percent in a matter of two to three weeks. Growth of wood structures has not been examined in many cases. Also, patterns of growth of new roots in fruit trees are not consistent, even from year to year under the same trees. It appears that root production only occurs when the environment (temperature and water) and internal competition allow. This suggests that roots are very weak competitors in spite of their obvious importance. In stone fruit, a flush of root growth is often noted after fruit harvest near the end of the season.

Partitioning to Fruit

A common expression in crop physiology for the percentage of total dry matter partitioned to the harvested portion (the fruit in this case) is the "harvest index" (HI). In temperate tree fruit, HI will be zero in the early years before cropping, but it can become very high in mature trees. The HI values from our studies and those reported in the literature for pome and stone fruit trees have been summarized. In general, the HI increases over orchard development from planting to maturity. For apples, values of 30 to 50 percent seem common, but can be as high as 65 to 80 percent of dry matter produced each year (Figure C1.2). Although fewer studies have been done with stone fruit, peaches also are reported to reach HI values of up to 70 percent. Comparable studies have not been done with cherries, but their yields are typically about half those of peaches due in part to very short fruiting seasons. We would expect cherries to have somewhat lower maximum HI values.

Research values for apples and peaches are extremely high compared to most crops and may in fact be too high for sustained cropping. Nonetheless, the data indicate the great ability of these crops to produce fruit. The factors mentioned earlier of preexisting structure, long leaf area duration, and low respiration are likely key to high maximum HI values in temperate fruit trees. Most annual field crops have HI values of 20 to 50 percent, excluding root systems (whereas the fruit tree values quoted previously included roots).

Partitioning to Vegetative Organs

As the HI increases with onset of cropping or different levels of crop at any stage, there are concurrent declines in partitioning to other organs. All vegetative organs receive fewer carbohydrates, but the greatest relative reductions are in root systems (Figures C1.1 and C1.2), followed by shoots, stems, and leaves. Leaf areas are reduced somewhat, as increases in crop load tend to reduce shoot growth, although amounts vary among studies. Partitioning to wood structures tends to decline rapidly with initial low yields, but then levels out to

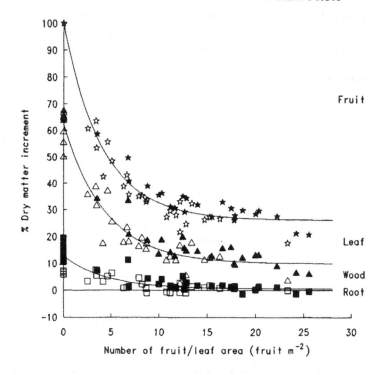

FIGURE C1.2. Effects of increasing crop load on partitioning of dry weight in dwarf apple trees (*Source:* Reproduced with permission from Palmer, 1992, *Tree Physiol.* 11:19-33.)

more constant amounts with higher crop loads. Apparent baseline amounts of leaf and structure partitioning are likely due to early season vegetative growth that occurs before fruit become strong competitors. Also, minimum quantities of leaf area and shoot structure are required for setting large crops of fruit.

In the case of roots, several studies in apples and peaches indicate that net seasonal dry weight gain may be near zero with the heaviest crop loads. This may be deceiving, however, since some of the carbohydrates partitioned to root systems are lost due to death of fine roots after only a few weeks or months. Trees can have fairly high root turnover rates, and new roots may be functional for a period of only a few weeks. Ignoring these roots is analogous to not including the leaves that fall each autumn in the partitioning to the top of the tree.

Current studies indicate that only a small percentage of the new roots produced each year will survive to become larger structural roots that we can easily find and measure. Nonetheless, cropping dramatically reduces the partitioning of carbohydrates to permanent roots in temperate fruit crops.

Partitioning to and from Carbohydrate Reserves

Another important process that requires carbohydrates is the accumulation of reserves. These are not as easily measured or interpreted. The nonstructural carbohydrate reserves are usually stored as polysaccharides such as starch or hemicellulose. They provide the critical carbohydrate supply for early season flower and shoot development before leaves can begin new photosynthesis. The general seasonal pattern of carbohydrate reserves is a maximum in early winter, followed by some decline with early bud development before visible growth and a rapid decline to a minimum shortly after bloom. Then, there is a gradual increase in reserves as the season progresses, reaching a maximum again as leaves fall.

For pome fruit, such as apple, the bloom minimum in reserves occurs about one month after budbreak when trees have about 20 percent of their leaf area. About 30 to 50 percent of extractable reserves are utilized before recovery begins. A tree can then support fruit growth with current photosynthesis. In stone fruit, however, bloom and budbreak occur at about the same time, the pattern changing with latitude and year. Therefore, reserve carbohydrates are needed for a longer period to support fruit growth until the leaf canopy can develop. In cherries, four to six leaves per shoot must be formed before shoots begin to contribute carbohydrates back to a tree. Also, it appears that at the minimum, stone fruit carbohydrate reserves are almost fully utilized before recovery. The partitioning of carbohydrate reserves in the spring is primarily to growing shoot tips and flowers, and to new root tips (area of first growth). Many of the carbohydrates are used for respiration, as early growth is intensive and requires a lot of energy.

Partitioning to Flower Bud Development

Flower buds are developing for subsequent years at the same time fruit and other organs are growing. Although partitioning of carbohydrates to these tiny buds is necessary, the amount is extremely small, and it is not clear if the carbohydrate supply is ever limiting. However, fruit load and effective leaf area seem to affect the process of flower bud initiation. Early thinning will promote flower bud initiation, while early leaf fall will decrease it. There is evidence that flower bud initiation or triggering is hormonally controlled, but there is also evidence that general carbohydrate relations may be involved as well. Some research suggests the initial triggering of flower bud development may be hormonally controlled, but the subsequent bud development that determines flower quality the next year may be carbohydrate related. The relative importance of the two factors remains to be clarified.

ENVIRONMENTAL FACTORS AFFECTING PARTITIONING

Light Availability

In the field, the amount of sunlight varies greatly over the day and season as well as across climates. Several studies, mostly with apples, have examined carbohydrate partitioning as affected by light level at specific times or over a full season. Based on these studies with apples and cherries, it appears that low light levels reduce carbohydrate production of nonfruiting trees and thus tend to reduce partitioning to roots compared to leaves and shoots. This may be interpreted as a plant response to a reduced need for water and nutrients from the soil and a need to grow shoots to find more light (called "etiolation" in the extreme case).

With fruiting trees, the effects can be complex. If fruit set has already occurred, low light reduces fruit growth and especially vegetative growth, as the fruit compete strongly for carbohydrates. If, however, low light occurs early in the season before fruit are set, growing shoots tend to be stronger competitors and fruit set may be reduced, even leading to complete defruiting. Then, partitioning develops as just described for nonfruiting trees.

Water Availability

Water deficits often limit productivity in orchards. The effects of water stress on carbohydrate partitioning are primarily on growth of the top of a tree, while the dry matter partitioned to roots is generally more stable. This could be due to the roots experiencing less water stress than the shoots, which must draw water through the plant. It would also help establish a better functional balance of water-absorbing roots to transpiring shoots when the limiting resource is obtained by roots. This behavior has been utilized to develop an irrigation strategy suited for stone fruit in arid climates, in which irrigation is withheld in midseason if shoots are too actively growing. This inhibits shoot growth for easier management but will not affect fruit growth very much if it is done at the time of the midseason slow growth phase of stone fruit mentioned previously. This is a good example of the utilization of plant physiology studies for practical crop management.

Nutrient Availability

Variations in mineral nutrient availability are relatively common in spite of excellent work that has defined the requirements and fertilization for temperate fruit trees. In general, it appears that if nutrients are limiting, the growth of the top of the tree is reduced more than the growth of the roots (similar to the response to water deficits). In some studies, roots actually grow more with low nutrient availability, presumably to obtain more of the limiting nutrients. Conversely, with high nutrient availability, top growth is stimulated while root growth may be reduced. When nitrogen is too high, flower bud initiation is inhibited and a tree can become too vegetative.

Pruning and Canopy Management

Pruning is always a dwarfing process. The earlier the pruning, the greater the effect in the current season. There are many practices that growers use to regulate growth and sustained cropping of temperate fruit crops by directly or indirectly affecting carbon partitioning. The most direct is the removal of young fruitlets (called "thinning") to reduce competition so that the remaining fruit can grow larger and have better quality. The reduction in crop load of course also allows more carbohydrate partitioning to other organs and processes such as flower bud development.

Pruning and training techniques (bending or positioning of branches, etc.) also are used to structure trees and to maintain good sunlight distribution in tree canopies. Pruning reduces the number of growing points and therefore tends to increase vigor of remaining shoots. This tends to reduce the number of fruiting sites and lower crop loads, which will affect partitioning and make trees more vegetative.

The partitioning of carbohydrates produced by photosynthesis is a critical process in the growth and cropping of temperate fruit trees. It is a complex whole-plant process that is a function of the unique combination of competing organs of the tree and is greatly influenced by many internal and environmental factors. Many cultural practices are essentially designed to regulate carbohydrate partitioning for optimizing yields of quality fruit in a given season and in the following years.

Related Topics: FLOWER BUD FORMATION, POLLINATION, AND FRUIT SET; FRUIT GROWTH PATTERNS; LIGHT INTERCEPTION AND PHOTOSYNTHESIS; PLANT GROWTH REGULATION; PLANT NUTRITION; WATER RELATIONS

SELECTED BIBLIOGRAPHY

American Society of Horticultural Sciences Colloquium (1999). Carbohydrate economy of horticultural crops, Colloquium proceedings. *HortScience* 34:1013-1047.

Flore, J. A. and D. R. Layne (1996). "Prunus." In Zamski, E. and A. A. Schaffer (Eds.), *Photoassimilate distribution in plants and crops* (pp. 825-849). New York: Marcel Dekker.

Lakso, A. N., J. N. Wunsche, J. W. Palmer, and L. Corelli-Grappadelli (1999). Measurement and modeling of carbon balance of the apple tree. *HortScience* 34:1040-1047.

Palmer, J. W. (1992). Effects of varying crop load on photosynthesis, dry matter production and partitioning of Crispin/M.27 apple trees. *Tree Physiol.* 11:19-33.

– 2 –

Cultivar Selection

Duane W. Greene

Each year, chance seedlings are found in orchards, hedgerows, and fields, and thousands of crosses are made by fruit breeders. Some seedlings have obvious flaws that make their elimination relatively easy. However, a large number of chance seedlings and controlled crosses are evaluated further to identify the few cultivars that will have broad consumer appeal. A new fruit selection must have a basic set of attributes in order to become a new and important commercial cultivar. Specific criteria and the importance of individual traits may vary with region or country. This chapter discusses the criteria used in identifying the very small select group of temperate fruit that will or may become commercially important.

SELECTION CRITERIA

Size

Medium- to large-size fruit are preferred and are the most profitable. For example, apple cultivars that have the majority of fruit in the 7 to 8 centimeter size category are favored in the marketplace (Hampson and Quamme, 2000). Cultivars that are genetically inclined to produce fruit in the smaller size categories are less profitable, since additional expense will be required to adjust culture to produce larger fruit. Fruit can be too large, however. Cultural adjustments to reduce size on large-fruited cultivars add expense and may reduce quality and increase the possibility of biennial bearing.

Appearance

The initial selection of a pome or stone fruit for further evaluation is usually due to attractive appearance. Color is a principle attribute. Apples that have a clear, glossy yellow or a bright red or green color will have an advantage. Peaches are preferred that have a bright red blush over a large portion of the surface. Cherries with a deep red color are generally more desirable. Fruit shape is another consideration. Generally, an apple with a moderate to high length to diameter (L/D) ratio is preferred to a very flat fruit. A round peach is usually more desirable than an oblong or very flat one. Surface blemishes, such as russet or scarf-skin on apples, detract from appearance and customer appeal.

Flavor

Flavor is emerging as being the most important attribute that a fruit must have to become commercially successful. Aroma and complex flavors are favored and sought after by many consumers. A recent survey of customers in three major cities in the United States reveals that over 90 percent of consumers consider taste to be the most important factor when making apple purchasing decisions (Ricks, Heinze, and Beggs, 1995). Consumers have become accustomed to having a number of alternative flavors available in the produce section, and a fruit that lacks the flavor component will not sell and will not be commercially viable for growers to plant.

Flavor has two aspects: sweet/tart ratio and aromatics. Distinct preferences exist among people. These differences can manifest themselves as national, regional, or local preferences. Fruit that are predominately sweet and have relatively low acids are favored by Asian populations in the Pacific Rim area. Apple cultivars such as 'Fuji' or 'Tsugaru' are preferred and grown in that region. Individuals who live in northern Europe generally prefer fruit that are more tart and have higher acid levels. Apples grown there include 'Jonagold', 'Cox's Orange Pippin', and 'Elstar'. New peach cultivars offer consumers more flavor choices than in the past. Some of the recent selections have low acid contents. The majority of consumers in the world favor an appropriate balance between sweet and tart.

Identification of eating quality of fruit using analytical methods such as firmness, soluble solids, or titrateable acidity is relatively unreliable and therefore of limited use in a screening process (Watada et al., 1981). Increasingly, sensory evaluation of fruit is being used as a method to identify and select promising new cultivars. Taste panels have proven to be quite reliable predictors of preferences consumers have for different cultivars and strains of cultivars.

Texture, Firmness, and Juiciness

Texture, firmness, and juiciness are attributes that are intimately associated with, but distinctly different from, flavor. Barritt (2001) suggests that crispness is the most sought after trait with apples. Flesh should literally crack, and upon chewing, the apple should melt and disappear quickly. Ricks, Heinze, and Beggs (1995) also report that crispness is a key factor consumers identify as being important in making apple purchase decisions. Peaches are preferred that have smooth, melting flesh. Cherries should be crisp yet mellow. Associated with texture is juiciness. This is a highly sought after trait that is equated with freshness. A fruit must be firm, but, above all, it must be juicy and crisp or melting to be appealing enough to purchase in preference to other cultivars or produce. These attributes must remain relatively unchanged for a number of days after the consumer takes the fruit home.

Yield and Culture

A new cultivar must have characteristics that will allow the grower to make a fair profit. It should have attributes that will allow high annual yields with substantial percentage packouts in the highest fruit grades. Some favorable cultural characteristics include annual flowering, ease of thinning, high productivity, absence of physiological problems or nutrient disorders, and desirable tree growth.

Storage and Shipping Quality

Bruising is one of the most common reasons that temperate fruit are downgraded or culled. Cultivars that do not bruise easily are preferred. Customers consistently select pome or stone fruit that maintain firmness and texture for an extended period of time and are not subject to storage disorders. Therefore, handling and storage characteristics are important criteria used to identify exceptional selections.

Disease Resistance

Tree fruit traditionally have required many sprays to produce commercial, blemish-free crops. Many of these sprays are applied to control diseases. Resistance to the most common and devastating diseases can be found in a number of apple cultivars. Apple scab *(Venturia inaequalis)* is the most serious disease of apples worldwide. Popular cultivars such as 'Liberty', 'Enterprise', and 'Gold Rush' are scab resistant. Bacterial spot is considered a serious disease on peaches, and its control requires multiple fungicide sprays. Peach and nectarine cultivars are selected for reduced susceptibility to this disease.

Cold Hardiness

In areas that have severe winters, production is limited by winter injury or even death of trees. Cultivar selection is based on the ability of trees not only to withstand extremely cold temperatures but also to survive under fluctuating temperature conditions. The newly introduced apple cultivars from the University of Minnesota, 'Honeycrisp' and 'Zestar', illustrate that selection for cold hardiness does not necessarily come at the expense of selection for high quality. 'Madison', 'Harcrest', and 'Canadian Harmony' are examples of peaches that were selected for both quality and winter hardiness.

Winter Chilling Requirement

Production of temperate fruit in warm, subtropical areas is limited because most of the cultivars available require too many hours of chilling. In areas where production is limited due to low chilling, early leafing is a screening condition used to identify selections that may grow with little or no chilling.

Miscellaneous Criteria

Selection criteria vary by region, based on local conditions. For example, a grower-sanctioned program in Ohio selects mainly for lateness in bloom, a trait that may help avoid spring frosts prevalent in that area. Other selection and evaluation factors include flesh color, flesh oxidation, tree growth habit, tree vigor, and season of ripening.

MULTIDISCIPLINARY AND INTERNATIONAL EVALUATION OF CULTIVARS

A regional project joining over 50 scientists in the United States and Canada was established in 1994, specifically to evaluate new apple cultivars. Promising new cultivars are evaluated for insect and disease susceptibility, horticultural characteristics, and organoleptic qualities over a wide range of climatic conditions. The goals of the project are to identify new superior cultivars and regions or climates that are most ideally suited to grow the better cultivars. Nearly 50 selections from breeding programs, chance seedlings, and mutations have been or currently are under evaluation. To date, no such regional projects have been organized for other tree fruit.

MAJOR CULTIVARS OF THE WORLD

Thousands of fruit cultivars exist throughout temperate growing regions. Some are only of local interest, while others are planted worldwide. Stone fruit cultivars generally have relatively narrow regions of adaptability, so there are no dominant, widely planted cultivars. In contrast, pome fruit, especially apples, have much wider ranges of adaptability, so internationally important apple cultivars do exist (see Box C2.1).

It is quite likely that cultivar development and introduction practices will change in the future. Many of the varieties that are most widely planted in the world are many years old, and fruit performance in marketing channels is only fair to good. Consumers have a plethora of choices of fruit and vegetables in supermarket produce sections, and they are becoming more discriminating about their purchases. Cultivars that are extensively planted in the future must be flavorful, fresh, and possess high internal quality, or consumers will purchase better tasting alternatives.

Related Topics: BREEDING AND MOLECULAR GENETICS; FLOWER BUD FORMATION, POLLINATION, AND FRUIT SET; HIGH-DENSITY ORCHARDS; MARKETING; ORCHARD PLANNING AND SITE PREPARATION; PLANT-PEST RELATIONSHIPS AND THE ORCHARD ECOSYSTEM; PROPAGATION; TRAINING SYSTEMS

BOX C2.1. Top Ten Apple Cultivars in the World

Dozens of apple cultivars are grown throughout the world, but only a few account for the majority of the total world production. The following, in descending order, were the top ten apple cultivars in the world in 2000, which together comprised over 70 percent of all apple production in the world (O'Rourke, 2000).

'Delicious'. Originally named 'Hawkeye', this was a chance seedling discovered in Peru, Iowa, by Jesse Haitt. Stark Brothers Nursery introduced this cultivar in 1895 with the name 'Delicious'. The fruit is medium-size and conic, with prominent protruding calyx lobes. The tree is productive and bears annually if thinned. 'Delicious' is hardy to zone 5. Over the years, many red-coloring sports and trees with spur-type growth habits have been identified, patented, and extensively planted. Although fruit from these trees is more attractive than the original 'Hawkeye Delicious', eating quality is widely acknowledged to be inferior.

'Golden Delicious'. 'Golden Delicious' originated as a chance seedling in the orchard of Anderson Mullins in Clay County, West Virginia. It was named and introduced by Stark Brothers Nursery in 1916. 'Golden Delicious' fruit is yellow, medium-size, and conic. It frequently develops russet when grown in humid climates. The tree is very productive and very grower friendly. 'Golden Delicious' is considered to be a high-quality apple and has served as a parent to many of the promising new cultivars.

'Granny Smith'. 'Granny Smith' was a chance seedling, with one of the parents thought to be 'French Crab'. It was discovered in Australia by Marie Ann (Granny) Smith and known to exist in her yard in 1868. It is a green, very late-maturing apple with very good quality and storage potential. Flesh is white and tart but becomes sweet in storage. The tree is extremely productive and possesses a growth habit that is easy to manage.

'Gala'. 'Gala' originated from a cross between 'Golden Delicious' and 'Kidd's Orange Red' made by J. H. Kidd in 1934. It was named 'Gala' in 1950. 'Gala' is considered an extremely high-quality apple that is medium in size. Several harvests are required, and as it approaches maturity, cracks can develop in the pedicel end. It was not planted heavily until the 1980s. Gala is revered for its excellent fruity taste.

'Fuji'. 'Fuji' originated from a cross of 'Red Delicious' and 'Ralls Janet' at the Tohoku Research Station, Japan, in 1939. It is a good-quality, medium-size, pink to light red apple that is sweet and stores very well. Fuji is popular in Japan and gained international recognition in the 1980s.

(continued)

(continued)

'Jonagold'. This apple resulted from a cross between 'Golden Delicious' and 'Jonathan' at the New York Agricultural Experiment Station in Geneva, New York. It was named by Roger Way and introduced in 1968. The fruit is large, conic, blushed pinkish red with extremely high quality. It is a triploid, so its pollen is not viable as a pollinizer. Fruit quality tends to be best in cooler climates. Storage life is medium to short, and it is quite susceptible to the calcium deficiency disorder that manifests itself as bitter pit.

'Idared'. This cultivar was selected in 1935 from a cross of 'Jonathan' and 'Wagner' made by Leif Verner at the Idaho Experiment Station. The fruit is midseason, medium to large, red, and round to conic, with white flesh. Flavor is mild, and quality is average. 'Idared' can be stored longer in regular atmosphere storage than most apples. It is considered both a dessert and a processing apple. 'Idared' is one of the earliest-blooming cultivars. The tree is moderate size and very grower friendly.

'Jonathan'. 'Esopus Spitzenburg' is believed to be the seed parent of this apple that was discovered before 1926 in Woodstock, New York. 'Jonathan' is a mostly red, small to medium, round dessert apple with whitish flesh. It ripens midseason and has a very mild flavor that is somewhat acidic. The tree is small and is noted for its susceptibility to mildew.

'Rome Beauty'. 'Rome' was a seedling discovered in Rome township, Ohio, in about 1832. It is a very large, burgundy red, late-ripening apple. The skin is thick and tough, the flesh is white, and the taste is subdued and mildly acidic. 'Rome' is a good cooking apple, but the dessert quality is fair at best. It stores very well in regular atmosphere storage.

'McIntosh'. This apple, believed to be a seedling of 'Fameuse', was discovered by John McIntosh in 1796. It is a midseason, medium-size, good-quality red dessert apple. 'McIntosh' frequently displays excessive preharvest drop, has soft flesh, and poor red color development. Consequently, it is grown commercially only in areas that are cool during the harvest period. The tree has above-average cold hardiness.

SELECTED BIBLIOGRAPHY

Barritt, B. H. (2001). Apple quality for consumers. *Compact Fruit Tree* 34:54-56.

Childers, N. E. and W. B. Sherman, eds. (1988). *The peach: World cultivars to marketing,* Third edition. Gainsville, FL: Horticultural Publications.

Greene, D. W. and W. R. Autio (1993). Comparison of tree growth, fruit characteristics, and fruit quality of five 'Gala' apple strains. *Fruit Var. J.* 47:103-109.

Hampson, C. R. and H. A. Quamme (2000). Use of preference testing to identify tolerance limits for visual attributes in apple breeding. *HortScience* 35:921-924.

O'Rourke, D. (2000). World apples to 2010. *The World Apple Report* 8(1):6-8.

Ricks, D., K. Heinze, and J. Beggs (1995). Consumer preference information related to Michigan apples. *The Fruit Grower News* 35:38-39.

Stebbins, R. L., A. A. Duncan, O. C. Compton, and D. Duncan (1991). Taste ratings of new apple cultivars. *Fruit Var. J.* 45:37-44.

Watada, A. E., J. A. Abbott, R. E. Hardenburg, and W. Luby (1981). Relationships of apple sensory attributes to head space volatiles, soluble solids, and titrateable acids. *J. Amer. Soc. Hort. Sci.* 106:130-132.

Webster, A. D. and N. E. Looney, eds. (1996). *Cherries: Crop physiology, production and uses.* Wallingford, UK: CAB International.

1. DISEASES
2. DORMANCY AND ACCLIMATION
3. DWARFING

– 1 –

Diseases

David A. Rosenberger

More than 350 diseases are known to affect temperate zone tree fruit. However, the majority of economic losses are attributable to fewer than 50 diseases, most of which are caused by fungi or bacteria. Viruses account for more than 50 percent of the known pome and stone fruit diseases, but many are relatively uncommon. Nematodes are the direct cause of a few diseases, and they contribute to others by vectoring viruses or by predisposing trees to pathogen attack. Diseases can also be caused by abiotic factors such as plant malnutrition, cold injury, or oxygen deprivation in water-logged soils.

Comprehensive descriptions of diseases and disease management programs are available in the publications cited at the end of this chapter. Management strategies vary significantly from one geographic region to another because of differences in climate, cultivars, pest complexes, and market objectives for the crops involved. Local and state extension systems can often provide the best regionally adjusted recommendations for disease control.

BACKGROUND ON TREE FRUIT DISEASES AND THEIR CONTROL

Diseases develop when a pathogenic organism encounters a susceptible host in an environment that allows infection to occur. Diseases can be controlled by eliminating any one of the three factors required to complete a disease cycle: pathogen, susceptible host, or conducive environment. Pathogen exclusion is the most common method used to control virus diseases in tree fruit and is often accomplished by selecting and planting only virus-free trees. For other diseases, host susceptibility is eliminated or minimized by using disease-resistant germplasm or by protecting otherwise susceptible tissue with fungicides or bactericides. Most fungal and bacterial diseases are dependent on rainfall for dissemination and infection, so environmental conditions required for disease development sometimes can be avoided by careful selection of planting sites and cultural management practices that hasten drying after dews and rains.

Integrated pest management (IPM) systems use multiple approaches to minimize pest damage. IPM systems seek to affect the pathogen, the host, and the environment in ways that minimize opportunities for diseases to become established. For example, an IPM approach for controlling fire blight in apples should include alternating small blocks of highly susceptible cultivars with blocks of more resistant apple cultivars so as to minimize potential spread of inoculum within orchards. If infections occur, inoculum for the next year can be reduced by pruning out infections as they appear and by removing cankers during winter. Host susceptibility is reduced by avoiding excessive nitrogen fertilization and by spraying trees during bloom to prevent primary infections of blossoms. Using integrated approaches for managing difficult diseases is usually more effective than depending on a single practice. However, where effective fungicides are available, fungicides alone may provide the most cost-effective way of controlling fungal diseases.

Diseases are easiest to manage if control measures are applied before the pathogen becomes well established in the orchard. Thus, most control measures are aimed at controlling the initial infections in spring. With many fungi and bacteria, pathogen numbers can escalate exponentially with each reproductive cycle. Sanitation measures or pesticide applications that are effective when applied early in the season may be only partially effective when applied to a population in the log phase of epidemic development.

Careful timing of pesticide applications is essential for cost-effective control, especially early in the growing season. Pathogen levels and availability of inoculum cannot be easily assessed with any visual methods, so timing of controls must be based on tree phenology, weather conditions, or models that predict optimum control timing. Later in the growing season, timing of fungicide cover sprays is often adjusted to coincide with insecticide spray timing so as to minimize application costs by controlling both insects and diseases with a single trip through the orchard.

Fungicides are available for controlling most fungal diseases of apples and stone fruit. Although fungicides have been used for many years, fungicide strategies are constantly evolving and changing as new fungicides are introduced and older products are discontinued. New cultivars often have different disease susceptibility patterns than the cultivars that they replace. And new discoveries in plant and pathogen biology continuously contribute to improvements in disease control strategies.

Bacterial diseases occur more sporadically than the major fungal diseases. Bacterial diseases are notoriously difficult to control because bacteria reproduce rapidly when conditions favor their development, and few bactericides are registered for treating fruit trees. Controlling bacterial diseases usually requires a combination of pathogen exclusion, sanitation measures to reduce inoculum levels where the pathogen is already present, and carefully timed antibiotic sprays to protect plants during periods of peak susceptibility.

Viral diseases can be prevented only by exclusion. No pesticides are available for controlling virus diseases in plants.

Many diseases cannot be identified with certainty based on field symptoms alone because various pathogens produce similar symptoms. The pattern of disease occurrence within a tree or an orchard sometimes provides clues that can help with diagnosis. However, accurate disease diagnosis is often possible only by isolating the pathogen into pure culture and identifying it via microscopic examination in the laboratory. This is especially true for pathogens causing summer fruit rots on apples, postharvest decays, cankers, and nondescript leaf spots.

COMMON POME FRUIT DISEASES

The major pome fruit diseases are listed alphabetically using commonly accepted disease names. Each disease name is followed by the Latin binomial for the pathogen(s), a letter to indicate whether the pathogen is a bacterium (B) or a fungus (F), and a brief description of disease symptoms and control strategies.

Apple replant disease is caused by an undefined complex of soil organisms. The disease causes stunting or reduced growth of young apple trees planted into old orchard sites. Various studies have implicated nematodes, actinomycetes, fungal root pathogens such as *Pythium,* herbicide residues from previous plantings, low soil pH, and drought stress of newly planted trees. Soil fumigation prior to planting is beneficial in some sites but not in others. Effects of the disease can be avoided at many sites by using good preplant site preparation coupled with irrigation of young trees during the year of planting.

Apple scab (F: *Venturia inaequalis*) is the most common and economically important disease of apples. Lesions appear on leaves and fruit nine to 17 days after infection (Figure D1.1), with the length of the incubation period depending on temperature. Young lesions appear as circular, velvety, olive-brown spots that gradually turn black-brown with age. Severely affected trees defoliate by midsummer and may fail to form fruit buds for the next year's crop. Fruit that become infected early in the season are misshapen and may crack. Infections on fruit pedicels may cause fruit to drop during June.

Primary inoculum consists of ascospores that are released from overwintering leaves on the orchard floor during rains. Ascospore release begins at or soon after trees reach budbreak in spring. It peaks when trees reach the pink- or full-bloom stage and terminates near petal fall or within two weeks thereafter. The "Mill's Table" lists the minimum wetting period required for infections to occur at various

FIGURE D1.1. Apple scab lesions on a leaf (left) and on fruit (right)

temperatures and is widely used to monitor potential infection periods. Infections initiated by ascospores produce abundant conidia that are disseminated to new leaves and fruit by splashing and wind-blown rain. Most infections on fruit occur as a result of secondary spread from primary lesions established on prebloom foliage. Infected fruit that have no symptoms at harvest may develop pinpoint scab (small black lesions 1 to 3 millimeters in diameter) during cold storage.

Scab is usually controlled by applying fungicides to prevent primary infections. Scab-resistant cultivars are available but are not widely accepted or easily marketed. Removing leaf litter before spring rains or using cultivation, urea sprays, or saprophytic fungi to speed leaf degradation can reduce the amount of primary inoculum but rarely provide complete control of scab.

Bitter rot (F: *Colletotrichum* species) is a common summer fruit rot in hot humid climates, but it is rare in cooler growing regions and in arid climates. Infections are most common on the side of fruit facing the sun. They start as sunken tan lesions that gradually enlarge, sometimes developing masses of slimy pink spores in the centers of lesions. Decayed tissue extends in a narrow "V" pattern toward the seed cavity. The disease can spread rapidly during hot, wet weather and is controlled with fungicides. Captan is often used to control bitter rot during summer.

Black rot (F: *Botryosphaeria obtusa*) is a summer fruit rot characterized by firm dark lesions, sometimes with a bull's-eye pattern, and sometimes with black pycnidia evident in the center of larger lesions. Infections on fruit can occur anytime from before bloom until harvest. Killed fruitlets that remain attached to the tree after fruit thinning are frequently colonized and provide inoculum both for later summer infections and for infections of fruit and foliage the next spring. Dead branches, including branches killed by fire blight, are rapidly colonized and will produce both ascospores and conidia if left in the tree over winter. The pathogen also causes frogeye leaf spot and limb cankers. Benzimidazoles, mancozeb, metiram, and strobilurin fungicides provide good control.

Blue mold (F: *Penicillium expansum*) is the most economically important postharvest decay of apples and pears. Decayed fruit flesh is soft and watery, separates easily from healthy tissue, and has a musty, earthy odor. Spores are disseminated by air or in water flumes. The fungus rapidly invades wounds in fruit and can also invade through fruit stems during long-term controlled atmosphere storage. Blue mold is controlled by using sanitation measures to limit exposure to inoculum and by using postharvest fungicide treatments.

Bull's-eye rot (F: *Pezicula malicorticis*) is a late-summer fruit decay that is common in northwestern United States and is occasionally found in more humid climates. Decay lesions are brown, often with a lighter brown or tan center. Affected tissue is slightly sunken, firm, and does not easily separate from healthy adjacent tissue. Inoculum comes from cankers in apple trees or from fungal colonies in the dead outer bark of pears. The disease is controlled with fungicides applied during late summer.

Cankers (various causes) are diseased areas on tree limbs or trunks where pathogens have killed the tree bark. Cankers can be caused by bacteria (e.g., fire blight cankers) or fungi. Common canker-causing fungi include *Botryosphaeria obtusa* (black rot canker), *Nectria galligena* (European apple tree canker), *Pezicula malicorticis* (anthracnose canker), and *Neofabraea perennans* (perennial canker). Many canker-causing fungi are weak pathogens that invade through wounds or stressed tissue. For example, black rot cankers on apples in northeastern United States occur primarily on limbs where xylem-inhabiting basidiomycetes such as *Trametes versicolor* or *Schizophyllum commune* have extensively colonized the internal woody cylinder within the limbs.

Cedar apple rust (F: *Gymnosporangium juniperi-virginianae*) causes yellow or orange lesions on leaves and fruit of susceptible apple cultivars in regions where the alternate host, eastern red cedar *(Juniperus virginiana),* is endemic. Basidiospores produced on galls in cedar trees are released during spring and early summer rains. Most infections on apple leaves and fruit occur between tight cluster and first cover, but leaf infections may continue for up to six weeks after petal fall. There is no secondary disease cycle on apples. Aeciospores produced on apple can only infect cedars. Fungicides are used to control this disease, but disease pressure can be reduced by removing cedars within several hundred feet of orchard perimeters.

Fabraea leaf and fruit spot (F: *Fabraea maculata*) is common where susceptible pear cultivars (e.g., 'Bosc') are grown in warm, humid climates. The disease appears during summer as small, 1 to 3 millimeter red or purple spots on leaves and fruit. The spots turn brown with age and produce an abundance of splash-dispersed conidia. The disease spreads extremely rapidly and can cause complete defoliation and crop loss. In northeastern United States, fungicides are needed from petal fall through mid-July to prevent primary infections. Later sprays are needed when primary infections are not completely controlled.

Fire blight (B: *Erwinia amylovora*) is the most destructive disease of apples and pears because it spreads rapidly and can kill entire trees. Bacteria released from cankers are spread to blossoms by insects and splashing rain. Blossom infections result in collapse of the blossom cluster and invasion of the subtending branch. On twigs killed by fire blight, the leaves collapse and turn blackish brown but do not abscise. Shoot tips on killed twigs frequently bend over to form a diagnostic "shepherd's crook." Bacterial ooze is usually visible on recently invaded leaves, twigs, branches, and fruit. Secondary shoot infections occur when inoculum contacts and invades succulent tissue in growing shoot tips. Rootstock blight occurs when the pathogen kills susceptible apple rootstocks even though the scion portion of the tree may be only mildly affected by fire blight. Under some conditions, the pathogen can move internally from blossom and twig infections to rootstocks without causing apparent damage to intervening tissue in limbs and trunks. Trauma blight occurs when bacteria invade tissue that is damaged by hail or frost. Apple and pear cultivars show wide variations in their susceptibility to fire blight.

Fire blight is controlled by pruning out infected tissue to reduce inoculum. A delayed dormant copper spray can help to suppress inoculum levels in orchards that were infected the previous year. The most important control measure involves well-timed applications of antibiotics (streptomycin, or terramycin where bacteria are resistant to streptomycin) during bloom.

Flyspeck (F: *Schizothyrium pomi*) appears during summer as a skin blemish on apples grown in warm, wet climates. Fungal colonies are usually 1 to 3 centimeters in diameter and consist of a few to more than 50 shiny black spots reminiscent of deposits left by a large fly (Figure D1.2). Most infections originate with ascospores or conidia blown into orchards from the numerous wild hosts found in hedgerows and wood lots. Infected fruit must be exposed to more than 250 hours of accumulated wetting before symptoms become visible. Fly-

FIGURE D1.2. Flyspeck (FS), sooty blotch (SB), and white rot (WR) on apple fruit at harvest

speck is controlled by application of fungicides beginning at petal fall.

Frogeye leaf spot: See black rot.

Gray mold (F: *Botrytis cinerea*) is the most important postharvest disease on pears and the second most important on apples. The decay is pale brown, soft, and watery on ripe apples and pears, but it can cause a firm decay of apples during controlled atmosphere storage. Conidia can infect through stems on pears or through wounds on pears and apples. Calyx infections that occur in the field may remain quiescent until fruit are moved to storage. *Botrytis cinerea* spreads rapidly from one fruit to other contacting fruit. As a result, infection originating in a single fruit can cause large losses during long-term storage.

Leaf spot (various causes) is a generic term often used for non-descript brown spots 1 to 3 millimeters in diameter. The spots can be caused by various fungi or by abiotic factors such as spray injury. Two common leaf spot diseases, frogeye leaf spot on apple and Fabraea leaf spot on pear, were described earlier under their own names. Cedar apple rust can cause extensive rust-induced leaf spotting in apple leaves where development of rust lesions is arrested by host resistance after rust basidiospores have already germinated and killed leaf cells. This rust-induced leaf spotting is common in 'Empire' and 'Cortland' trees that are not protected with fungicides during rust infection periods. Many weak pathogens, including species of *Alternaria, Phomopsis,* and *Botryosphaeria,* will invade leaf tissue that has been injured by rust or by other abiotic factors, but these pathogens have only limited capabilities for invading healthy leaf tissue. An exception is the strain of *Alternaria mali* that causes Alternaria leaf spot in southeastern United States and in Asia.

Pear blast (B: *Pseudomonas syringae*) affects pear flowers during cool, wet weather and can result in significant reductions in fruit set. Symptoms include blackening of the calyx ends of fruit and, occasionally, collapse of entire spurs. However, the subtending wood beneath killed spurs remains healthy. Losses to this disease are often blamed on light frost or poor pollination. Streptomycin sprays applied prior to infection periods have been shown to reduce losses.

Pear scab (F: *Venturia pirina*) on pears parallels apple scab on apples. The life cycle and control measures for pear scab are very similar to those described for apple scab.

Phytophthora crown and root rot (F: *Phytophthora cactorum* and other *Phytophthora* species) is the most important soilborne disease of tree fruit. The pathogens release flagellated zoospores that can infect tree roots and crowns in water-saturated soils. The fungus kills the bark on the crown and larger roots. Aboveground symptoms become apparent after trees are girdled by the fungal infection. Infected tissue is usually apparent several inches below the soil line where infected bark and inner phloem tissue have a soft texture and rusty red-brown color when cut. The disease is best managed by planting trees on well-drained sites, tiling poorly drained sites prior to planting, planting trees on raised berms, and avoiding susceptible rootstocks such as MM.106 and M.26 on sites with questionable internal soil drainage. The fungicide mefenoxam can be used to protect susceptible trees.

Powdery mildew (F: *Podosphaera leucotricha*) is second only to apple scab as the most important fungal disease of apples. The fungus overwinters in infected buds and grows to cover the new green tissue when these buds begin growing in spring. Infected leaves develop a white powdery coating of spores and mycelium. The conidia from these primary infections can initiate secondary infections on new leaves. Fruit infected at the pink-bud stage develop a netlike russeting, but mildew is primarily a foliar disease.

Powdery mildew is the only aboveground fungal disease of tree fruit that thrives in dry climates because it does not require free water (rains or dews) for infection. Conidia can germinate on leaf surfaces when relative humidity is as low as 70 percent. This disease is controlled with fungicides. Some apple cultivars are relatively resistant to infection. Highly susceptible cultivars will require more fungicide protection than less susceptible cultivars.

Quince rust (F: *Gymnosporangium clavipes*) is a disease that affects apple fruit but not foliage. Infected fruit are often misshapen with deeply sunken lesions. The life cycle of quince rust is similar to that of cedar apple rust, but quince rust produces perennial cankers in cedars instead of galls.

Sooty blotch (F: a complex of *Peltaster fruticola, Leptodontium elatius,* and *Geastrumia polystigmatis*) causes superficial gray, black, or cloudy areas on apple fruit (Figure D1.2). The disease cycle and controls are similar to those described for flyspeck.

Virus diseases are less important in pome fruit than in stone fruit. Many old apple cultivars and rootstocks contained one or more latent viruses such as apple mosaic virus, apple stem pitting virus, apple stem grooving virus, or apple chlorotic leaf spot virus. These viruses caused no visible symptoms on most cultivars and were considered benign even though some of them reduced productivity of some cultivars. Other viruses or viruslike diseases such as stony pit in pears cause fruit deformities that make fruit from infected trees unmarketable. Tomato ringspot virus causes a tree decline in certain cultivar-rootstock combinations (e.g., 'Delicious' on MM.106), but it does not affect most apple rootstocks.

The best defense against pome fruit viruses is to establish orchards using trees that are certified to be free of known viruses. Almost all of the pome fruit viruses are disseminated primarily via virus-infected propagation material. Tomato ringspot virus is the exception. It is vectored by several species of *Xiphenema* nematodes.

White rot (F: *Botryosphaeria dothidea*) is a summer fruit rot that occurs primarily in warm, humid growing regions. Lesions usually become visible during late summer as small, brown to tan spots, often with a red halo around the margins. As the decayed area expands (Figure D1.2), it extends in a cylindrical pattern toward the core of the fruit. (This contrasts with the "V"-shaped pattern characteristic of bitter rot.) Under warm conditions, rotten fruit have a soft, watery, "applesauce-in-a-bag" composition, but under cooler conditions white rot is not easily distinguished from black rot. The disease is controlled with fungicides (benzimidazoles, mancozeb, metiram, strob-ilurins) and by pruning out dead wood that can harbor this fungus.

COMMON STONE FRUIT DISEASES

Bacterial canker (B: *Pseudomonas syringae* pv. *syringae*) affects all stone fruit but can be especially severe on sweet cherries and apricots where it causes cankers and kills spurs. The bacteria overwinter in cankers, buds, and sometimes in symptomless host tissue. Severe infections of cherry leaves and fruit often occur in association with light frosts. Symptoms consist of irregular spots on leaves and sunken lesions on fruit. Cankers on twigs and branches are often initiated in autumn. Copper sprays applied at leaf fall have been used to control the canker phase in sweet cherries, but some strains of the pathogen are resistant to copper.

Bacterial spot (B: *Xanthomonas arboricola* pv. *pruni,* formerly *Xanthomonas campestris* pv. *pruni*) causes spots on leaves, fruit, and twigs of peaches, nectarines, Japanese plums, and apricots. In the United States, it occurs east of the Rocky Mountains and is especially severe where trees are grown on light, sandy soils in humid environments. The bacteria invade leaf scars in autumn and overwinter in buds, small cankers, or twig surfaces. Bacteria spread to new leaves and fruit during rains beginning at late bloom. Spots on leaves originate as angular purple lesions 1 to 3 millimeters in diameter and are usually concentrated along the leaf midrib or toward the tips of leaves. Spots on fruit may become visible about three to five weeks after petal fall. The disease is best controlled by selecting resistant cultivars. Susceptible cultivars must be protected with copper sprays in fall and/or spring to reduce inoculum levels and terramycin or copper at petal fall and early cover sprays to protect fruit and leaves.

Black knot (F: *Apiosporina morbosa*) appears as black swellings or knots on twigs and branches of plums and cherries in eastern United States. Ascospores are released from the knots during spring rains beginning when trees are at white bud and continuing for several weeks after petal fall. The ascospores infect nodes on new shoots. New knots usually become apparent about one year after infection and produce ascospores the second year. Severely affected trees become unproductive. Black knot is controlled by removing infected wild *Prunus* from hedgerows prior to planting, by pruning out infected knots as they appear, and by protecting trees with fungicides during the period of ascospore release.

Brown rot (F: *Monilinia fructicola, M. laxa, M. fructigena*) is the most widespread and economically important fungal disease of stone fruit. It causes a blossom blight, twig blight, canker, ripe fruit rot, and postharvest fruit rot. *Monilinia fructicola* is the primary pathogen in eastern North America, whereas *M. laxa* predominates in Europe. Both species are found in California and South America. The brown rot fungi overwinter in cankers, mummified fruit in trees, and fallen fruit on the orchard floor. The latter may produce apothecia and ascospores in spring, but ascospores are usually less important in the disease cycle than conidia that are produced on cankers or mummies in trees. Blossoms and fruit become infected during warm rains. Green fruit are more resistant to infection than blossoms or ripening fruit. However, green fruit may develop quiescent infections that become active as the fruit ripen in the field or after harvest. Brown rot is controlled by pruning out cankers and fruit mummies during winter and by using fungicides to protect blossoms, ripening fruit, and fruit after harvest.

Cherry leaf spot (F: *Blumeriella jaapii*) affects both sweet and sour cherries in eastern North America and Europe. Ascospores are produced in apothecia on overwintering leaf litter and are released during rains starting during late bloom and continuing for about six weeks thereafter. Infections appear as small red to purple spots on the upper leaf surface that gradually enlarge to 3 millimeters and turn brown. On the underside of leaves, lesions appear pink or cream-colored during wet weather due to production of conidia. Secondary infections continue to occur through summer and can cause early defoliation, thereby leaving affected trees susceptible to winter damage. The disease is controlled by applying fungicides to prevent primary infection.

Leucostoma canker or Cytospora canker (F: *Leucostoma persoonii*) affects peaches, nectarines, and sweet cherries and is especially severe in regions where trees may be damaged by cold winter temperatures. Cankers on older wood are usually elliptical, blackened, and gummy. Conidia are produced in cankers throughout the year. Infections occur only through wounds, dead tissue, or injuries (including sunburn and cold injury). The disease can be avoided by keeping new plantings away from old diseased plantings and by using good horticultural practices to minimize tree stress and injuries. Fungicides are not effective.

Peach scab (F: *Cladosporium carpophilum*) is important in warm humid production regions. It affects primarily peaches and nectarines, but can also occur on plums and apricots. The disease overwinters in infected buds and twigs. Conidia are released during periods of high humidity beginning about two weeks after shuck split and can infect fruit, leaves, and green twigs. Fruit infections first appear as small pinpoint, olive-green spots that gradually enlarge to 2 to 3 millimeters in diameter. The disease is controlled by applying fungicides for several weeks after shuck split.

Peach leaf curl (F: *Taphrina deformans*) causes red blotching, thickening, and puckering of peach and nectarine leaves. It occasionally affects fruit. The fungus overwinters as a yeastlike saprophyte on twigs and in buds, then invades developing leaf tissue during cool, wet weather in early spring. It can be controlled with copper sprays or other fungicides applied at leaf fall in autumn or before budbreak in spring.

Phytophthora crown and root rot (F: *Phytophthora* species): See comments under Common Pome Fruit Diseases.

Powdery mildew (F: *Sphaerotheca pannosa, Podosphaera clandestina, P. tridactyla*) can affect all stone fruit and is most severe where stone fruit are grown in arid climates. Infected leaves and shoots develop a powdery white coating of mycelia. *Sphaerotheca pannosa* overwinters in buds on peaches and roses, with the latter sometimes acting as an inoculum source for orchard infections. *Podosphaera clandestina* overwinters as cleistothecia trapped in the bark of cherry trees or on the orchard floor. Ascospores are released from cleistothecia during spring rains, but conidia from primary infections can spread and infect during periods of high relative humidity in the absence of rain. The greatest losses occur on sweet cherries, when young fruit become infected and deformed. The disease is controlled with fungicides applied to prevent early season infections.

Virus or viruslike diseases cause extensive losses in stone fruit. The most common viruses are prunus necrotic ringspot virus (PNRSV) and prune dwarf virus (PDV). The numerous strains of PNRSV cause a variety of symptoms, the most common being a necrotic leaf spot that appears the first year a tree is infected but rarely thereafter. Infected trees may show no obvious symptoms, but productivity is often reduced. Both PNRSV and PDV can be transmitted by seed and by pollen. Pollen transmission allows the disease to spread rapidly from tree to tree in the field. PDV causes sour cherry yellows, a disease characterized by leaf yellowing and abscission in midsummer and presence of long, barren shoots caused by a lack of fruiting spurs. PNRSV and PDV are controlled by using virus-certified planting stock and by keeping new plantings away from old infected orchards.

Tomato ringspot virus (TmRSV) causes constriction disease of plums and prunus stem pitting in peaches, nectarines, and cherries. The disease is most common in temperate growing regions of the eastern United States. The virus is seed transmitted in dandelion and presumably in other weed hosts. It is also transmitted by dagger nematodes (*Xiphinema* species). European plums propagated on *Myrobalan* rootstocks develop a constriction below the graft union and a brown line or pitting in wood at the graft union. Affected trees decline rapidly four to seven years after trees are planted. On peaches, nectarines, and cherries, TmRSV causes deep pitting in the woody cylinder of the rootstock, and affected trees decline. Control measures include preplant soil treatments to reduce or eliminate populations of vector nematodes and regular use of broadleaf herbicides to keep alternate hosts from becoming established in the orchard.

Plum pox, or sharka, is caused by plum pox virus (PPV) and is common throughout Europe. PPV is vectored by aphids. Depending on the species and cultivar of stone fruit, PPV may cause deformed fruit, loss of productivity, tree decline, or no visible symptoms at all. PPV was recently introduced in Pennsylvania and Canada but is being controlled by eradication of infected trees.

X-disease is caused by a phytoplasma and affects peaches, nectarines, and sweet cherries. Phytoplasmas lack cell walls, live in plant phloem, and are vectored by leafhoppers. Leafhoppers acquire the phytoplasma from infected sweet cherry trees or from wild hosts, primarily chokecherry *(Prunus virginianae)* or naturalized sweet cherry seedlings. Leaves on affected peach and nectarine limbs develop red, water-soaked lesions and abscise prematurely, leaving a tuft of young leaves at the ends of denuded shoots. The disease usually kills peach, nectarine, and cherry trees on *P. mahaleb* rootstock within two to four years. Cherry trees on mazzard rootstock *(P. avium)* survive many years and can be detected only by uneven fruit ripening and production of small fruit with delayed maturity. The disease can be controlled only by eliminating infected hosts within 500 feet of new plantings.

A wide array of fungal, bacterial, and viral pathogens can attack fruit, leaves, wood, and roots of temperate zone fruit trees. Viruses cause the largest number of diseases, but most viral diseases are rare in commercial orchards. Fungal diseases are common and would cause total crop loss in many years and locations if they were not controlled with fungicides. Bacterial diseases, although few in number, are difficult to control and cause extensive losses in some years. Researchers continue to devise cost-effective IPM strategies to manage the common diseases of tree fruit.

Related Topics: INSECTS AND MITES; NEMATODES; PLANT-PEST RELATIONSHIPS AND THE ORCHARD ECOSYSTEM; SUSTAINABLE ORCHARDING

SELECTED BIBLIOGRAPHY

Agnello, A., J. Kovach, J. Nyrop, H. Reissig, D. Rosenberger, and W. Wilcox (1999). *Apple IPM: A guide for sampling and managing major apple pests in New York State,* Pub. 207. Geneva, NY: New York State IPM Program.

Hogmire, H. W. Jr., ed. (1995). *Mid-Atlantic orchard monitoring guide,* Pub. NRAES-75. Ithaca, NY: Northeast Regional Agric. Engin. Serv.

Jones, A. L. and H. S. Aldwinckle (1990). *Compendium of apple and pear diseases.* St. Paul, MN: APS Press.

Németh, M. (1986). *Virus, mycoplasma and rickettsia diseases of fruit trees.* Dordrecht, the Netherlands: Martinus Nijhoff Publishers.

Ogawa, J. M. and H. English (1991). *Diseases of temperate zone tree fruit and nut crops,* Pub. 3345. Oakland, CA: Univ. of California, Div. of Agric. and Nat. Resources.

Ogawa, J. M., E. I. Zehr, G. W. Bird, D. F. Ritchie, K. Uriu, and J. K. Uyemoto (1995). *Compendium of stone fruit diseases.* St. Paul, MN: APS Press.

Ohlendorf, B. L. P. (1999). *Integrated pest management for apples and pears,* Second edition, Pub. 3340. Oakland, CA: Univ. of California, Div. of Agric. and Nat. Resources.

Solymar, B., M. Appleby, P. Goodwin, P. Hagerman, L. Huffman, K. Schooley, A. Verhagen, A. Verhallen, G. Walker, and K. Wilson (1999). *Integrated pest management for Ontario apple orchards,* Pub. 310. Toronto, Ontario: Ontario Apple Marketing Commission.

Strand, L. L. (1999). *Integrated pest management for stone fruits,* Pub. 3389. Oakland, CA: Univ. of California, Div. of Agric. and Nat. Resources.

– 2 –

Dormancy and Acclimation

Curt R. Rom

Dormancy is a condition in which plants or plant parts are alive but not growing. Dormant plants—also called quiescent, latent, asleep, or suspended—are not visibly developing, and there is no visible external activity. Deciduous fruit trees are considered dormant during the winter season after leaf fall and during other periods of environmental stress. Tissues that express dormancy are apical and root meristems, lateral and axillary meristems, and cambial and cork meristems. Organs that become dormant are buds, root tips, and seeds. Dormancy is common in all temperate fruit trees, as may be inferred because of seasonal environmental variations. A plant physiologically changes, or acclimates, in response to its environment to ensure its continued existence.

Fruit trees express dormancy at different times of year as a survival tool to prevent growth during unfavorable conditions. For instance, when the temperature is too hot or too cold, tissues or organs may become dormant. Additionally, environmental factors such as light or water stress, either in excess or limiting, may cause plant parts to become dormant. Then, when more favorable conditions prevail, tissues are released from dormancy and begin growth, as indicated by cell division and expansion. Thus, plant parts, whole plants, or seeds may survive from season to season, flourishing in conditions favorable for growth.

FORMS OF DORMANCY

A number of terms have been used to describe types of dormancy (e.g., "quiescence," "rest"), what controls dormancy (e.g., "mechanical dormancy," "physiological dormancy," "correlative inhibition"), and the duration of dormancy (e.g., "shallow dormancy," "deep dormancy"). These terms indicate the complexity of dormancy, but they can be confusing. Universal terminology, introduced by G. A. Lang and others (1985, 1987), provides a good model for explaining and understanding dormancy (Table D2.1). Three terms are used to describe the fundamental forms of dormancy. "Ecodormancy" is dormancy of growing tissues and organs imposed by factors in the surrounding environment. "Paradormancy" is used to explain dormancy imposed by growth factors outside of a tissue or organ (i.e., the control one part of a plant exerts over the growth of another part of the plant). "Endodormancy" is dormancy imposed by growth factors from inside a dormant plant structure.

Ecodormancy occurs when environmental conditions are not suitable for growth. All meristems and meristematic organs express this form of dormancy. Thermal ecodormancy occurs when the temperature for growth is too low or too high. Hydrational ecodormancy occurs when available soil moisture limits growth or flooding causes growth to cease. Atmospheric ecodormancy and photoecodormancy occur when a gas (e.g., oxygen) and light, respectively, are limiting. Once an optimum growth environment returns, growth resumes.

Paradormancy, also referred to as correlative dormancy, is regulated by physiological factors outside a tissue or organ. The best ex-

TABLE D2.1. Definitions and examples of three fundamental forms of plant dormancy

Type of Dormancy	Definition	Examples
Ecodormancy	Regulated by *environmental* factors	Temperature extremes Water stress Nutrient deficiencies
Paradormancy	Regulated by *physiological* factors outside the affected structure	Apical dominance Protective seed coverings
Endodormancy	Regulated by *physiological* factors inside the affected structure	Chilling requirement Physiological seed dormancy Photoperiodic response

Source: Modified from Lang et al., 1985.

ample in fruit trees is the phenomenon known as apical dominance, in which an actively growing apex suppresses growth of subtending buds on the same shoot. Following removal of the apical meristem by pruning or pinching, paradormancy ends, allowing lateral and axillary buds to begin to grow.

Endodormancy is one of the more complex forms of dormancy. In response to changing environmental conditions (e.g., shorter days, lower average daily temperature, lower internal cell moisture content), a bud or seed enters a state of prolonged dormancy. Unlike ecodormancy, growth will not automatically resume when there is an appropriate growing environment. Thus, it is often referred to as deep dormancy or rest. Endodormancy is regulated by physiological factors inside a plant structure. For instance, internal biological mechanisms sense passage of time based on temperature. After exposure to specific cool temperatures for specific periods of time, endodormancy is eliminated, or rest is satisfied, and growth may once again resume. During endodormancy, plant parts acclimate to the external changing environment and may gain or lose "hardiness," depending upon both the external conditions and the state of dormancy.

SPECIFIC CASES OF DORMANCY

Seed Dormancy

A seed is, in essence, an embryonic dormant plant awaiting appropriate conditions to grow. Seed dormancy allows for seedling survival, increased dispersal of the seed, and synchronized germination of seeds.

Seed dormancy can occur in several forms. Ecodormancy is a controlling factor. In simple seed dormancy, if a seed does not have sufficient water for imbibition, optimum temperature conditions, or an appropriate light and gaseous environment, it will not germinate and grow, although alive and carrying on metabolic activity at a low rate.

Some seeds have evolved physical, paradormancy mechanisms. For instance, certain seeds have hard, sclerified seed coats (testa) or other outer coverings that are not easily penetrated by water or gases. Dormancy control is not removed until the seed coat or surrounding tissue has been broken, scarred, exposed to digestive acids of animals that aid in their dispersal, decomposed either chemically by soil acids or biologically by soil microorganisms, or physically removed. For horticultural purposes, some seeds require scarification—physical scarring by abrasives or organic acids, or cracking and removal. For example, a peach "pit" is actually a sclerified, bony endocarp tissue containing a seed. This protective pit prevents rapid water entry and must be removed, cracked, or chemically dissolved in order for the seed inside to germinate and grow.

Seeds of most temperate zone fruit also express endodormancy; they must be exposed to a period of cool, moist conditions prior to germination. The cool, moist requirement is called "stratification." Stratification may be a relatively short period (days to weeks) or repeated cycles of cooling followed by warm weather (repeated years) to ensure that seeds last over several winter seasons. For nursery operators and propagators, knowledge of the stratification requirements of specific seeds is essential if a high degree of germination is to be achieved. Seeds can have multiple dormancy mechanisms to ensure their survival and dispersal. It is common that ecodormancy of seeds overlaps their endodormancy, so that even though the endodormancy requirements are met after the appropriate stratification period, they will not germinate and grow until the appropriate environment for that growth is apparent to the seed.

Bud Dormancy

Fruit tree buds also express the three basic forms of dormancy. Ecodormancy is apparent in many woody plants. Early in the season, when temperatures are mild and sunshine is abundant, there is a flush of vigorous vegetative growth. Later in the season, if conditions become too hot or dry, growth will slow and may cease, and a terminal bud may be set at the apex of a shoot. However, if favorable conditions return, the plant produces a new flush of growth. Sometimes, several cycles of growth are observed in woody plants as they respond to changes in temperature, water availability, and light during the growing season.

Below the apex of a shoot, lateral buds may express paradormancy, as mentioned previously. Tree fruit vary in their degree of apical dominance and thus the manifestation of paradormancy. Apples, pears, and sweet cherries typically exhibit strong paradormancy of lateral buds and strong dominance of the apex. Thus, heading-back pruning cuts are used to stimulate the development of lateral branching. Many stone fruit, particularly peaches and tart cherries, have weaker apical dominance and tend to more readily develop lateral branches below the apex. Paradormancy appears to be a stronger influence on vegetative buds than on floral buds.

In temperate fruit trees, as daylength shortens and temperatures cool, buds express ecodormancy. With increasingly shorter daylength and lower temperatures, endodormancy, especially pronounced in floral buds, is then induced. Some reports indicate that the point in time that a bud changes from ecodormancy to endodormancy is a stage of vegetative maturity. It is from this point in time that there is an accumulation of cool temperature exposure, or "chill," by the plant. After a specific period of chill exposure, endodormancy ends. The period of chill required for buds to physiologically change and have endodormancy removed is referred to as "chilling requirement."

Various models have been proposed for predicting when fruit trees will complete endodormancy. A model for peach trees is based on the hourly accumulation of chill units (CUs) (Richardson, Seeley, and Walker, 1974). It is a partial sine wave or extended quadratic model in which 1 CU is accumulated when flower buds are exposed for 1 hour to a temperature of 7.2°C. No CUs are accumulated at temperatures below freezing or at 12.8°C, and there is a negative CU (–1 CU) response at 21°C. In 1990, Linvil introduced a modified model utilizing daily high and low temperatures. Researchers propose that the onset for the accumulation or vegetative maturity of chill begins after flower buds have received a threshold number of CU (approximately 50) without interruption by negative chill at high temperatures.

PHYSIOLOGICAL BASIS FOR DORMANCY

Just as the forms of dormancy are complex and diverse, the explanations for dormancy are varied. Simple models attribute ecodormancy to balances of water, carbohydrates, and energy within the growing meristem. As environmental conditions become less favorable, metabolic functions slow or change. For instance, as temperatures go below or above those that are optimal for metabolism, the reactions slow and thus growth slows or stops. Also, as tissues dehydrate or carbohydrate content lowers, metabolism may slow. Paradormancy is generally explained by a simple mechanistic model relating the balance of apical meristem-produced auxins and root-translocated cytokinins. Hormonal and nutritional models are used to explain endodormancy within buds and seeds. Bud scales and seed coats may contain the hormone abscisic acid (ABA). The tissue concentration of ABA is often highest as endodormancy begins and dissipates with the exposure of tissues to cool temperatures over time. More recently, biochemical and molecular studies are shedding some new light on these dormancy mechanisms and their controls from a genetic standpoint.

ACCLIMATION

Coincident with the beginning and duration of dormancy in many plant organs and tissues is an ability to withstand increasingly harsher environmental conditions. For instance, with gradual temperature increases, a bud beginning high-temperature-induced ecodormancy has an increased ability to withstand higher temperatures without damage to the meristem. This process of increased environmental tolerance with exposure is called acclimation. Typically, acclimation is a relatively slow process—it may take days to weeks of slowly changing environmental conditions for a tissue or organ to acclimate to an environmental extreme. Rapid changes can still cause damage.

The loss of acclimation is called deacclimation. As conditions revert to those that are optimum for growth, tissues may deacclimate, or lose their ability to withstand environmental extremes. Similar to acclimation, deacclimation is a relatively slow process, which prevents the plant from losing its ability to withstand harsh conditions if there is a brief interlude of favorable conditions.

During endodormancy, cold temperature acclimation is very important. Tissues and organs acclimating to cold temperatures are said to increase in hardiness—the ability to withstand very cold temperatures. Generally, as endodormancy begins and plants are exposed to shortening days and decreasing temperatures, the tissues acclimate to the cold temperatures and gain in hardiness. Within a given species, the rate of acclimation may relate to the rate of temperature decline. For some woody fruit trees, studies demonstrate that plants are most hardy as they near completion of the endodormancy requirement. As a result of this phenomenon, plants may be somewhat vulnerable to late-autumn, early winter freezes and less susceptible to midwinter freezes. Once endodormancy is complete, plants may deacclimate and lose hardiness with exposure to warmer temperatures and lengthening photoperiod. As the tissues are in postendodormancy ecodormancy, they retain some ability to regain hardiness within some reasonable temperature span. As growth begins after ecodormancy, hardiness fades and the tissues become very sensitive to freezing temperatures.

Knowledge of the causes, forms, and physiology of fruit tree dormancy and acclimation has improved in recent years. New research in the area of molecular biology has the potential to further increase understanding of these complex processes. However, a complete and simple story based upon molecular evidence has not yet unfolded.

Related Topics: BREEDING AND MOLECULAR GENETICS; GEOGRAPHIC CONSIDERATIONS; PLANT HORMONES; TEMPERATURE RELATIONS

SELECTED BIBLIOGRAPHY

Dennis, F. G. Jr. (1994). Dormancy: What we know (and don't know). *HortScience* 29:1249-1254.

Lang, G. A. (1987). Dormancy: A new universal terminology. *HortScience* 22:817-820.

Lang, G. A., ed. (1996). *Plant dormancy, physiology, biochemistry, and molecular biology.* Oxon, UK: CAB International.

Lang, G. A., J. D. Early, N. J. Arroyave, R. L. Darnell, G. C. Martin, and G. W. Stutte (1985). Dormancy: Toward a reduced, universal terminology. *HortScience* 20:809-811.

Linvil, D. E. (1990). Calculating chilling hours and chill units from daily maximum and minimum temperature observations. *HortScience* 25:14-16.

Richardson, E. A., S. D. Seeley, and D. R. Walker (1974). A model for estimating the completion of rest for Redhaven and Elberta peach trees. *HortScience* 82:302-306.

Silverton, J. (1999). Seed ecology, dormancy and germination: A modern synthesis from Baskin and Baskin. *Amer. J. Bot.* 86:903-905.

Viémont, J. D. and J. Crabbé, eds. (2000). *Dormancy in plants: From whole plant behaviour to cellular control.* Cambridge, UK: Univ. Press.

– 3 –

Dwarfing

Stephen S. Miller

The culture of dwarf tree fruit dates from early times. By definition, a dwarf plant is one that is smaller than normal size at full maturity. A dwarf tree usually has other characteristics in addition to stunted growth or a reduced stature. For example, precocity, canopy architecture, time of flowering, and fruit size may be altered in the dwarf tree, although tree size is certainly the predominate character and the one most often associated with genetically based or culturally induced dwarf trees.

Interest in dwarf trees is based on the many advantages they offer in orchard management and enhanced fruit quality. Due to smaller stature, dwarf trees provide labor savings in pruning, harvesting, and spray application. Light penetration is generally greater in dwarf tree canopies, which improves photosynthesis and fruit quality. Discussions on orchard management of dwarf fruit trees are found in other parts of this book.

From an anatomical standpoint, dwarf trees are not different from standard trees, but horticultural practices can alter the physiology of a tree, resulting in dwarfing. What causes a tree to be dwarfed and how can a fruit tree be manipulated to produce a dwarfed tree? In practice, dwarf trees may result from genetic changes (natural or imposed through select breeding) or through horticultural manipulations (e.g., rootstock, pruning, training, scoring, cropping, deficit irrigation, plant bioregulators, etc.). These methods will be briefly discussed with emphasis on the physiology of the dwarfing process.

GENETIC DWARFING

Within a large population of trees, a certain number of genetic dwarf variants will occur (Schmidt and Gruppe, 1988; Scorza, 1988), generally from less than 1 percent up to 2 or 3 percent of the total population. The mechanism(s) for dwarfing may be one or more of several inherent structural or physiological characters, such as a spreading growth habit, shortened internodes, a change in hormone levels, a tendency toward basitonic (from the base) growth, or decreased vigor. Despite significant advances in genetic engineering through biotechnology, the genes responsible for dwarfing have not been identified, although recent studies are providing a better understanding of the mechanisms that cause dwarfing. Mutations in nature can also result in dwarf trees. Spur strains of apple are a result of limb or whole-tree mutations. It has been suggested that high light intensity may induce these mutations, but there are no supporting data. Spur growth habit trees are often 25 to 50 percent smaller than standard trees from which they mutate.

HORTICULTURAL PRACTICES TO INDUCE DWARFING

Rootstocks and Interstocks

The use of a rootstock or interstock to dwarf a scion variety is an age-old technique. Grafting a vigorous cultivar (the scion) on the top of a certain tree (the stock) is known to produce various degrees of dwarfing. The degree of dwarfing achieved depends on the rootstock but may also be influenced by other factors, such as the natural vigor of the scion, the soil, and cultural practices. The rootstock may be the same species as the scion, as in apple, or a different species, as in pear (quince root). More research has been conducted on rootstocks and their effect on the scion than any other aspect of dwarfing. Still, the mechanisms for dwarfing by the rootstock are not well understood. In 1956 Beakbane stated, "we know much about the extent, something about the duration, but we still have some way to go before we can say that we fully understand the fundamental nature, or mechanism, of rootstock influence." Over 40 years later we have a better understanding of the dwarfing influence of rootstocks, but we still do not have a full understanding of how rootstocks dwarf the scion variety.

For certain, no one mechanism is responsible, and it is likely that several mechanisms, working concurrently, in tandem, or independently, are responsible.

One early theory suggested that the roots of a dwarfing rootstock occupy a smaller volume of soil, grow to less depth, and absorb less nitrogen than roots of a standard size rootstock, thereby resulting in a smaller scion. While the root system of a dwarf tree does occupy less soil volume than a vigorous seedling (the roots are in equilibrium with the top), research has failed to fully support this theory. Nutrient uptake, especially phosphorus, has been implicated as a growth-controlling mechanism in dwarf trees; however, results of numerous studies have been inconsistent in establishing nutrient uptake as a dwarfing mechanism. Evidence is available that demonstrates that restricted nutrient translocation across the graft union between rootstock and scion plays a role in the dwarfing mechanism. Anatomical changes within the xylem and other tissues at the graft union that affect the movement of nutrients and water from rootstock to scion have been implicated as a dwarfing mechanism (Simons, 1987). Hormones, especially auxins that are translocated from shoot tips to roots, are thought to act as "regulators" in nutrient and water flow across the graft union. Unfortunately, the extent of influence by these factors in the dwarfing process is unclear. Recent work by Atkinson and Else (2001) indicates that a gradient in hydraulic resistance exists within the rootstock union, ranging from high to low for dwarf, semidwarf, and vigorous rootstocks, respectively. Resistance to water flow at the graft union affects sap flow rate and the concentration of hormones and other solutes in the sap, which affects growth of the roots and shoots.

The primary plant hormones—auxins, gibberellins, and cytokinins—have been studied and implicated in the dwarfing mechanism. However, their precise role, alone or in combination, remains unclear. Auxins promote root growth, and studies show that dwarfing rootstocks have lower levels of auxin than more vigorous stocks. Auxin levels affect differentiation of xylem and phloem tissue, thus affecting the flow of nutrients, water, and assimilates (plant food). Gibberellins (GAs) affect cell elongation and have a major effect on shoot growth. There is no evidence that root-produced GAs influence shoot growth, and most researchers feel they play a minor role, if any, in the dwarf rootstock effect. Cytokinins, on the other hand, are produced in great quantities in the roots and translocated upward, where they influence shoot growth. Dwarfing rootstocks show high levels of cytokinins accumulating at the graft union, which indicates that rootstocks may affect growth by reducing upward movement of this vital growth hormone. Evidence has also been presented that dwarfing rootstocks have significantly higher levels of abscisic acid (ABA) than more vigorous stocks.

In most tree fruit that respond to grafting, dwarfing can be obtained by grafting a piece of stem from a dwarf rootstock between a more vigorous stock and a scion variety. This suggests that the mechanism for dwarfing is associated with the rootstock stem piece and not the root system. The longer the interstock stem piece or the higher a scion is budded on the rootstock, the greater the dwarfing effect. Much of this effect is attributed to bark phenols and their effect on auxin (indoleacetic acid) metabolism. It is hypothesized that as auxin moves basipetally, it is degraded and the concentration decreases. The thicker bark and much higher starch levels in dwarf rootstocks indicate a low level of auxin in these tissues, lending support to the bark phenol hypothesis.

Root Pruning and Root Restriction

Removing or pruning a portion of a tree's root system will temporarily reduce the top growth of the tree. The ancient art of bonsai is dependent on this technique of root pruning. Time of root pruning and distance from the tree trunk have an effect on the degree and length of growth control achieved. Root pruning in the spring near bloom has a greater dwarfing effect than root pruning in mid- or late summer. When a tree's roots are severed, growth of new roots is stimulated in the area of the cut. The tree's natural root:shoot equilibrium is upset, and, in response, assimilates are directed away from top growth to the roots. The result is a decrease in shoot growth. As new roots are produced, the root:shoot equilibrium is reestablished and the dwarfing effect is lost. When properly applied, the effect of root pruning may last an entire growing season. Other complex and interactive physiological mechanisms also contribute to the change in top growth when trees are root pruned. Absorption of water and nutrients is reduced, and hormone synthesis, particularly cytokinins, is decreased. A reduction in water absorption results in water stress, which reduces transpiration, causing stomatal closure and a reduction in photosynthesis. In combination, these effects contribute to reduced shoot growth. Because many of these physiological responses to root pruning are short-lived, root pruning may need to be repeated several times during a growing season to achieve effective growth control.

Dwarfing can also be achieved by restricting roots to a small area. This effect is readily observed in potted plants or in the culture of bonsai trees. In the field, hardpans restrict root growth to a shallow soil, resulting in dwarfed growth. In Australia, soil hardpans restrict root growth and allow apple trees to be grown at higher densities on vigorous rootstocks with reduced shoot growth. Planting trees at high density results in root competition and restricted horizontal spread of roots, a form of root restriction. Under these conditions, roots grow to greater depths in the soil. More recently, in-ground fabric containers have been used to restrict root systems, thereby reducing shoot growth. The physiological effects of root restriction have not been well studied in fruit trees. Unlike root pruning where root growth is stimulated, root restriction does not result in new root growth. When roots are restricted, root density increases, leaf transpiration and photosynthesis decrease, and foliar nutrient levels decline. These changes are probably responsible for reduced growth.

Dormant and Summer Pruning

Pruning is a dwarfing process and the degree of dwarfing is related to the severity of pruning. Light pruning may produce no recognized dwarfing response; the response to severe pruning may be easily seen and extend for several years. Pruning disturbs the balance between root and shoot growth. Root growth is slowed or stops as assimilates and growth hormones are directed to renewed shoot growth. Reduced root growth means less water and nutrient absorption. This furthers the stress and results in a dwarfing response by the tree. Dormant pruning disrupts apical dominance by removal of the auxins contained in the apical bud that prevent growth of the lower lateral buds. It is known that auxins are important in root growth. Loss of auxins through pruning affects root growth, which affects cytokinin production that is necessary for shoot growth. Summer pruning removes leaves, which affects tree water potential and photosynthesis. Summer pruning has long been regarded as more dwarfing than dormant pruning, but recent research does not support this belief.

Training

Bending a tree's branches from a vertical to a more horizontal position is a method of dwarfing. As a branch is reoriented by bending, terminal extension growth and apical dominance are reduced, and lateral branching is increased. Dwarfing resulting from limb bending is a hormone-controlled mechanism. When branches are bent, tissue is damaged and ethylene is released, which reduces terminal growth and increases lateral budbreak and diameter growth of the bent limb. Bending also affects auxin production and movement in the shoot. Less auxin is available to suppress lateral budbreak, and the auxin concentration is greater on the lower side of the bent limb than the upper side. The response to branch bending is greater in the lower part of a tree than in the upper part of the canopy. Cultivars respond differently to the degree of bending.

Scoring or Girdling

Scoring is cutting the bark around the circumference of a limb or the trunk down to the cambium layer. Girdling (also called ringing) is similar to scoring except a strip of bark is removed, usually about 6.4 millimeters wide, around the limb or trunk circumference. Both techniques interrupt the flow of carbohydrates to the root system, which slows root growth. Carbohydrates are diverted to buds. Reduced root growth slows movement of water, nutrients, and hormones to the shoots, thus affecting growth. The scored area produces new conducting tissues, reestablishing the connection between roots and shoots. Generally this occurs in midseason when terminal growth naturally slows or ceases. Scoring is used primarily in apple but recently has also been used in peach and nectarine to increase fruit size.

Deficit Irrigation

Withholding water at specific times when fruit growth is not adversely affected is termed regulated deficit irrigation and can be an effective dwarfing technique in arid growing areas. Research shows that regulated deficit irrigation reduces vegetative growth of peach, pear, and apple. A reduction in water availability affects tree water status or water potential (the free energy of water that is potentially available to do work relative to pure water) and results in water stress.

Water stress reduces turgor pressure that is directly related to cell expansion. Changes in turgor pressure also affect other physiological processes that lead to reduced growth. During periods of water stress, a greater percentage of assimilates are partitioned to roots, leaving less for shoot growth.

Cropping

Cropping will dwarf fruit trees. A heavy crop reduces shoot and root growth and leaf size. Photosynthetic efficiency is increased, but the increased efficiency is not enough to make up for the greater percentage of carbohydrates going to the fruit. Cropping also lowers the tree water potential, which reduces vegetative growth. Environment dictates, to some extent, which is more important in reducing growth—carbohydrate competition or reduced water potential. In high sunlight, low rainfall environments, reduced water potentials probably have a greater impact than carbohydrate partitioning; the opposite is likely in cloudy, high rainfall environments.

Plant Bioregulators (Plant Growth Regulators)

Application of selected plant bioregulators (PBRs) can inhibit growth in fruit trees (Miller, 1988). Auxins (NAA), ethylene-releasing compounds (ethephon), and GA biosynthesis inhibitors (paclobutrazol, prohexadione-calcium) are the primary PBRs used to reduce growth. Daminozide (Alar), a powerful growth regulator used for several decades in apple production, was removed from the market in the late 1980s. Research indicates daminozide reduced the translocation of GAs or GA precursors to actively growing sites and may have promoted GA catabolism and conjugation.

Auxins applied to scaffold limbs of trees induce bud dormancy, preventing growth of water sprouts from latent buds. Ethyl ester forms of auxins applied at high rates are phytotoxic and desiccate shoots. Ethylene is known to reduce cell or stem elongation. Compounds such as ethephon have been shown to interfere with auxin biosynthesis and with polar auxin transport. These auxin-mediated effects are probably the mechanism by which ethephon reduces shoot growth.

The most common group of PBRs used to control growth and induce dwarfing in fruit trees are those that interfere with GA biosynthesis. These retardants can be divided into three groups: quaternary ammonium compounds (e.g., chlormequat chloride), compounds with a nitrogen-containing heterocycle (e.g., flurprimidol, paclobutrazol, uniconazole), and acylcyclohexanediones (e.g., prohexadione-calcium). Each group interferes at a specific place in the GA biosynthesis pathway, inhibiting the endogenous formation of biologically active GAs. These active GAs (primarily GA_1) are significant in the longitudinal growth of plants. Blocking production of the active GAs results in shortened internodes, stem thickening, and darker green foliage—characteristics of a dwarf tree.

The dwarfing process for tree fruit is a complex response involving genetic and/or physiological changes that result in a tree of smaller stature. Although the specific mechanisms that lead to dwarfing are varied, depending on the horticultural technique(s) employed, and often not well understood, the outcome is a tree that is easier to manage and likely to be more efficient than a similar nondwarf tree.

Related Topics: BREEDING AND MOLECULAR GENETICS; CARBOHYDRATE PARTITIONING AND PLANT GROWTH; HIGH-DENSITY ORCHARDS; PLANT GROWTH REGULATION; PLANT NUTRITION; ROOTSTOCK SELECTION; TRAINING AND PRUNING PRINCIPLES; WATER RELATIONS

SELECTED BIBLIOGRAPHY

Atkinson, C. and M. Else (2001). Understanding how rootstocks dwarf fruit trees. *Compact Fruit Tree* 34:46-49.

Beakbane, A. B. (1956). Possible mechanisms of rootstock effect. *Ann. Appl. Biol.* 44:517-521.

Faust, M. (1989). *Physiology of temperate zone fruit trees.* New York: John Wiley and Sons.

Miller, S. S. (1988). Use of plant bioregulators in apple and pear culture. *Hort. Rev.* 10:309-401.

Schmidt, H. and W. Gruppe (1988). Breeding dwarfing rootstocks for sweet cherries. *HortScience* 23:112-114.

Scorza, R. (1988). Progress in the development of new peach tree growth habits. *Compact Fruit Tree* 21:92-98.

Simons, R. K. (1987). Compatibility and stock-scion interactions as related to dwarfing. In Rom, R.C. and R. F. Carlson (eds.), *Rootstocks for fruit crops* (pp. 79-106). New York: John Wiley and Sons.

Tukey, H. B. (1964). *Dwarfed fruit trees.* New York: The Macmillan Co.

1. FLOWER BUD FORMATION,
POLLINATION, AND FRUIT SET
2. FRUIT COLOR DEVELOPMENT
3. FRUIT GROWTH PATTERNS
4. FRUIT MATURITY

Flower Bud Formation, Pollination, and Fruit Set

Peter M. Hirst

Flowering and fruit development, from an evolutionary viewpoint, are mechanisms whereby a plant distributes seed and ensures propagation of the next generation. Conversely, the fruit grower has no concern for seed per se but is interested in the production of high yields of high-quality fruit. These goals tend to be somewhat conflicting, which represents a challenge for commercial orchardists.

FLOWER FORMATION

Fruit production starts with the initiation of a flower, pollination, and subsequent fertilization and fruit development. Seedling trees must go through a transition from a juvenile to adult stage before flowering can occur. Grafted trees, however, are adult above the graft from the time they are planted in the orchard, so juvenility is not a consideration in commercial fruit orchards.

In temperate fruit, flowers are formed within the buds during summer and fall, overwinter, and then bloom the following spring. In apple and pear, buds are either vegetative or mixed (containing leaves and flowers) with five to six flowers per bud and seven to eight flowers per bud, respectively. Flowers are typically borne on spurs, which are short shoots on two-year and older wood. Lateral buds on the previous year's growth may also produce flowers, depending on cultivar and growing environment. Apple flowers borne on the preceding year's growth open later and are generally smaller, producing smaller fruit. In peach and nectarine, however, flowers are borne laterally on one-year-old shoots. At each node, one to three buds may be borne, depending on cultivar and the vigor status of the tree. Where one bud is borne, it may be either a fruit or leaf bud. Where two are borne, one is usually fruitful while the other is not, and where three buds are present on a node, the center bud is often a leaf bud with a flower bud on each side.

Many factors affect whether a bud becomes reproductive or remains vegetative. Since flowers are initiated in the buds the year before they bear fruit, events one year can affect cropping the following year. The most obvious example of a year-to-year carryover effect on flowering is biennial or alternate bearing. This is fairly common in apple orchards and occurs when trees are overcropped one year, resulting in poor flower bud development and low crops the following year. For many years, this was thought to be because heavy crops depleted the tree of energy or nutrients, and the tree simply did not have enough energy to produce a sizeable crop the following year. In 1967, however, Chan and Cain published a classic paper demonstrating that the cause of poor flowering following a heavy crop is not heavy cropping per se, but rather the presence of seeds. Now it seems the gibberellins (GA) produced by developing seeds are the source of the inhibition of flower initiation, and some gibberellins (GA_7) appear more inhibitory to flowering than others (GA_4). Continued research examining the types, amounts, and transportation of various GAs in apple has failed to explain the differences among cultivars in their tendency for biennial bearing. The mechanisms controlling flower formation are complex, and although they have been widely studied, our understanding of the process is still fairly rudimentary. Although biennial bearing remains a problem in many orchards, judicious use of chemical thinning agents can help reduce its severity.

Flower initiation usually occurs during early spring, within approximately one month of the time of bloom. The first visible signs (Figure F1.1) of development occur when the flower begins to differentiate in the bud, usually during mid- to late summer. The transition of the apex of the bud from a flattened to a domed shape indicates that a bud has become floral rather than vegetative, although the invisible signal or switch obviously occurred some time earlier. A concept referred to as the "critical appendage number" embodies the idea that the bud must attain a certain degree of complexity for doming to occur. For apple, 18 to 22 appendages are required in spur buds prior

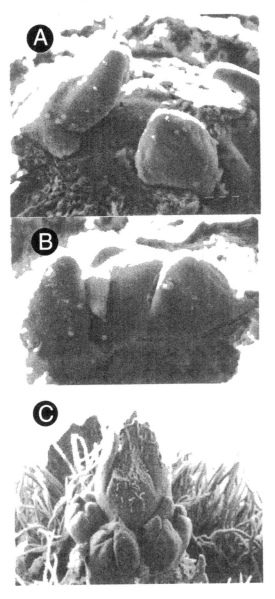

FIGURE F1.1. Electron micrographs of flower development in buds of 'Delicious' apple: (A) vegetative bud with flattened apex; (B) doming of the apex, the first visible sign of flower initiation; (C) flower bud of apple showing development of the king blossom and three lateral flowers to flower formation, although buds on previous year's growth flower at a low level of complexity.

Following doming of the apex, the flower parts (sepals, petals, anthers, etc.) differentiate in preparation for budbreak the following spring. Some evidence suggests that the degree of differentiation and/or flower size may play a role in the size of fruit produced, and studies are currently progressing to understand this further.

POLLINATION

Almost all temperate fruit species require pollination and seed set to produce commercial crops. A few cultivars of apples and pears are capable of setting fruit without seeds, while some species (such as *Malus hupehensis* and *Malus sikkimensis*) set seed apomictically; in other words, seed is produced vegetatively from a source other than the zygote. Some fruit crops require cross-pollination (apples, pears, most sweet cherries) while others are self-fertile (most peach cultivars, sour cherries, most apricots). European and Japanese plums vary in their requirements for cross-pollination by cultivar, with about half being self-fruitful. For apples, most cultivars will pollinate most others, with two key exceptions. Triploid cultivars, such as 'Mutsu' and 'Jonagold', produce nonviable pollen and therefore should not be considered pollinizers. Also, closely related cultivars (such as parents, siblings, or sports) may not work well as pollinizers for each other. 'Golden Delicious' has historically been regarded as having some degree of self-compatibility, but recent research shows the self-fertilization potency of this cultivar to be quite low. In addition to producing viable, compatible pollen, a pollinizer must flower at the same time, or slightly prior to, the cultivar for which it is intended to provide pollen. Typically, pollen is viable only for a short period of time, a matter of hours, and, therefore, synchronous flowering is important.

Successful pollination depends on both adequate and timely pollinizers (the source of the pollen) and pollinators (the agents of pollen transfer). Many nut trees rely on wind pollination, but the pollen of most fruit species tends to be heavy and not suited to wind pollination. As a result, pollinators are especially important. Pollination requirements of the crop should be considered at the time of orchard planning. For those fruit requiring cross-pollination, several approaches may be taken, but the most common is to avoid planting large blocks of a single cultivar. Bees tend to fly up and down rows rather than across rows, especially in orchards where trees form a continuous canopy rather than discrete trees. Orchard blocks of any one cultivar should be no more than five to six rows wide—this provides the best compromise between ensuring good pollen dispersal on one hand and efficient orchard management on the other. In some of the larger fruit-growing regions of the world, production is based mainly on just one or two cultivars, making large production areas of one cultivar desirable. To facilitate adequate pollination in such orchards, pollinizer trees can be planted throughout the orchard (usually every third tree in every third row). Crab apple species are frequently used for this purpose in apple orchards, with *Malus floribunda* and *Malus* 'Profusion' being popular choices due to their profuse flowering and commonality of flowering time with many commercial cultivars. Alternatives to pollinizer trees are hive inserts, which are packs of pollen that may be purchased and placed inside the bee hives so that bees are coated with suitable pollen upon leaving the hive, and bouquets of different cultivars placed in buckets of water and spaced around the orchard. Obviously, both hive inserts and bouquets require additional management time and are expensive to maintain and, consequently, are seldom used in commercial orchards.

The honeybee is the most prevalent pollinator for most fruit crops, although a number of other insects may play a secondary role. The level of pollination required differs among the various fruit crops, from a single-seeded fruit, such as a peach or a cherry, up to a kiwifruit with well over 1,000 seeds per fruit. Many factors determine the number of hives required to achieve optimal pollination, but two to eight hives per hectare is the generally accepted norm for tree fruit crops. Hives should be distributed at several locations around the orchard and introduced soon after the first flowers have begun to open.

FERTILIZATION

Pollination is not, of course, the final goal but is one step in the process ending in fertilization. Following the deposition of pollen to the stigmatic surface of a flower, a complex chain of biochemical recognition factors determines whether the pollen grain will germinate and also whether the pollen tube will grow down the stylar tissue of the flower toward the ovary. Barriers to self-pollination may occur at both these sites. Fertilization does not occur until the pollen tube has reached the ovule, but because the ovule is receptive for a limited time, the speed at which the pollen tube grows down the style becomes important. The effective pollination period (EPP) is the difference between the duration that the ovule is receptive and the length of time taken for the pollen to reach the ovule. The main factor determining pollen tube growth, and therefore EPP, is temperature. For example, in pear, the rate of pollen tube growth is more than five times faster at 15°C than at 5°C. Cool weather during the bloom period is detrimental, as it discourages bee flight. In addition, pollen tubes grow slowly during cool weather and may not reach the ovule while it is receptive. Although temperature is the primary determinant of pollen tube growth rate, other factors also play a role, such as the nutrient status of flowers, wind, and probably the light environment of the flowers.

Fruit set refers to the stage in which flowers are retained on the tree and develop into fruit, or else abscise. Shedding of flowers and young fruitlets occurs in several waves. During the first wave, unpollinated flowers are shed, followed by flowers pollinated but not fertilized. A number of fertilized flowers are shed in subsequent waves, depending on fruit species. In the Northern Hemisphere, this is called "June drop" ("December drop" in the Southern Hemisphere). Required flower set to ensure a reasonable crop obviously depends on the intensity of flowering but generally is in the range of 5 percent for apple up to as high as 70 percent for cherry. The degree of abscission is seldom sufficient to regulate crop load to attain good fruit size and return bloom, so fruit thinning is also required.

For multiseeded fruit, the number of seeds in retained fruit is important. Higher seed counts result in larger fruit size, although there appears to be a closer relationship between seed number and fruit weight in fruit with a high seed count, such as kiwifruit, than with lesser-seeded fruit, such as apple or pear. Nevertheless, even in fruit such as apple, it holds true that larger fruit on average have more seeds than smaller fruit. Seeds are also important for uniform fruit shape. Fruit with uneven seed distribution are often flattened or lopsided, with the side having fewer seeds being less well developed (Figure F1.2).

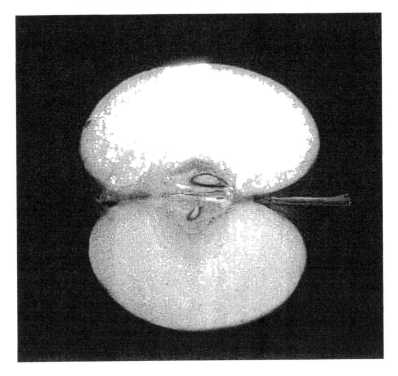

FIGURE F1.2. Uneven fruit shape caused by incomplete pollination. Note well-developed seed in the top half of the fruit compared with poorly developed seed and poorer fruit development in the lower half.

Flowering of fruit trees, from flower initiation to fruit set, is a complex process, and the details are not well understood. With most temperate tree fruit, however, horticulturists have quite comprehensive knowledge of the requirements of the crop and appropriate management techniques. As growers recognize, the foundation for a successful crop is laid the previous year when flowers are initiated. From this early beginning through the stages of pollination and fertilization, the challenge is to maximize the potential for high-quality fruit production.

Related Topics: ANATOMY AND TAXONOMY; ORCHARD PLANNING AND SITE PREPARATION; PLANT GROWTH REGULATION

SELECTED BIBLIOGRAPHY

Chan, B. and J. C. Cain (1967). Effect of seed formation on subsequent flowering of apple. *Proc. Amer. Soc. Hort. Sci.* 91:63-68.

DeGrandi-Hoffman, G. (1987). The honey bee pollination component of horticultural crop production systems. *Hort. Rev.* 9:237-272.

Fallahi, E. (1999). The riddle of regular cropping: The case for hormones, nutrition, exogenous bioregulators, and environmental factors. *HortTechnol.* 9:315-331.

2

Fruit Color Development

L. L. Creasy

The sensation of color is an interpretation by the brain of the reaction of retina in the eye to different wavelengths of electromagnetic radiation. If received in certain proportions, several combinations of radiation with wavelengths between 400 and 800 nanometers produce a sensation of white; any imbalance in these proportions produces color. A leaf looks green because it absorbs red and violet, leaving most of the yellow, green, and blue radiation that together add up to green. Red objects absorb energy at the blue end, leaving the red end of the radiation spectrum. Humans only see radiation between 420 and 800 nanometers, while other animals see other ranges.

Fruit colors cover the full visible spectrum from blue to red. The green color of plants is due to chlorophyll, a pigment essential for conversion of light energy into stored energy. Not surprisingly, the base color of most fruit is green, due to chlorophyll. Although some fruit stay green, others change color during ripening, a process that enhances their attractiveness to humans and to birds. Fruit color changes attract birds and occur after the seeds are mature enough to develop into new plants, resulting in effective seed dispersal. Frequently, ripening involves changes in sugar concentration and flesh texture, which also make the fruit more attractive to birds. Seed-eating birds appear to prefer black and red fruit, and many native fruit are these colors.

COLOR MEASUREMENT

Human perception is unrivaled in comparing both intensity and quality of fruit color. Nonsubjective methods are sought to document fruit color for quality standards and research purposes. The simplest method is the use of color comparison chips, which for many years provided a consistency in color description. Extraction and spectrophotometric quantification is accurate and reproducible but useful only in a research setting. A spectrophotometer measures absorption at selected parts of the electromagnetic spectrum in a chemical solution, which is a specific characteristic of the chemical. The method has the disadvantage of being destructive and therefore not applicable to all experimental designs. Spectroreflectometers (which measure the spectral energy reflecting from an object) can be employed and are not destructive but are generally immobile. Portable battery-powered color meters that rapidly and reproducibly generate color parameters for surfaces are useful in the field and are nondestructive. They have enabled many new research approaches in fruit color formation. The same technology expedited the development of rapid machine sorting by color.

PIGMENTS

Plant pigments have widely different chemical structures. The major chemical types are chlorophylls, carotenoids, anthocyanins, and betalains. The intensity of color is due to both the pigment concentrations in the fruit and their location within the fruit. Anthocyanins and carotenoids are antioxidants with potential health benefits to humans. Fruit extracts rich in them can be found in health stores in pill form.

The major pigment of fruit is chlorophyll, and the color of all green plant parts is due to this pigment. Carotenoids are lipophylic pigments that provide most of the yellow colors in nature, as also some red and orange. Most black, red, orange, blue, and purple coloration of fruit is due to the hydrophylic anthocyanins, glycosides of anthocyanidins. There are six major anthocyanidins (Figure F2.1), differing in substitution of the two rings, but most temperate fruit contain different glycosides of cyanidin. Another category of chemicals, which functionally and chemically replace anthocyanins in the Chenopodiflorae (including cacti and beets), is betalains (betacyanins and betaxanthins). Their presence results in plant colors similar to anthocyanins, but they have a totally different structure and biosynthetic origin. They do not occur in any temperate tree fruit.

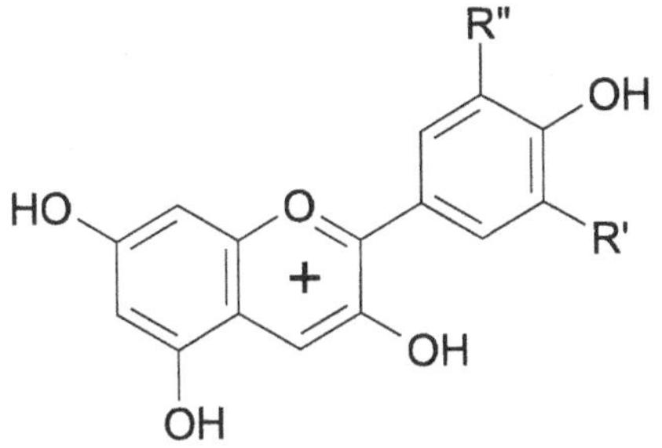

FIGURE F2.1. Anthocyanidins. The six most common are pelargonidin R"=H, R"=H; cyanidin, R'=OH, R"=H; peonidin R'=OMe, R"=H; delphinidin R'=OH, R"=OH; petunidin R'=OMe, R"=OH; and malvidin R'=OMe, R"=OMe.

Chlorophylls

Chlorophyll is a constituent pigment of plants, and many color changes in ripening fruit are due to a loss of chlorophyll unmasking other pigments. Light is necessary for chlorophyll synthesis in fruit. Chlorophyll is synthesized by a specialized pathway for tetrapyrrole rings. The starting point is the condensation of two molecules of gamma-aminolevulinic acid to form porphobilinogen. Four molecules of porphobilinogen condense to form a tetrapyrrole ring followed by subsequent modifications of the side chains, the incorporation of magnesium, and finally the attachment of the polyisoprene-derived long chain alcohol. Early steps in the biosynthetic sequence are similar to those in the biosynthesis of heme in animals; the process becomes plant specific at the insertion of magnesium.

Carotenoids

Carotenoids may be unmasked by a loss of chlorophyll but also may change in fruit during ripening. The most familiar change is the formation of the red pigment lycopene in ripening tomatoes. Similar changes occur in other fruit but have been less well studied. Carotenoids are synthesized starting with acetyl-CoA condensation to form mevalonic acid that is converted to isopentenyl pyrophosphate. Isopentenyl pyrophosphate condenses to form a C_{20} compound that further condenses to C_{40}, and cyclization results in a carotenoid. Many specific types of carotenoids are formed by isomeric modification, desaturation, cyclization, and introduction of oxygen molecules.

Anthocyanins

Anthocyanin pigments, although sometimes unmasked by loss of chlorophyll, are frequently synthesized from small precursors during the final stages of fruit maturation. Their appearance is subject to control by many environmental factors. The biosynthetic sequences for the different anthocyanidins are similar and represent one end-point of the flavonoid biosynthetic pathway, starting with the condensation of p-coumaryl-CoA with three malonyl-CoA units to form tetrahydroxychalcone that is then enzymatically isomerized to the flavanone naringenin. In the sequence to cyanidin, naringenin is converted to dihydrokaempferol, which is hydroxylated to produce dihydroquercetin. Further changes are less well documented, but it is thought that dihydroquercetin is reduced to a flavan-3,4-diol, and then oxidation, dehydration, and appropriate glycosylation produce the different cyanidin glycosides. The biosynthetic sequence is unique to higher plants and derived from the plant-specific production of the aromatic amino acids, phenylalanine and tryrosine.

Plants can metabolize anthocyanin molecules. A good example of this is chicory, in which new blue flowers open each morning and the anthocyanin is gone by early afternoon, leaving white flowers. In most fruit, synthesis and catabolism occur at the same time, and the concentration of pigment is a function of the synthesis rate and the catabolic rate.

The anthocyanin color of a fruit can be due to a single pigment (rare) or to mixtures of anthocyanins. In flower petals, it has been shown that two cultivars which can be distinguished by eye have identical chemical composition. Findings indicate that the pigments in one petal are mixed in the same cells, while in the other they are in different cell layers, causing unique light reflections. In many fruit, the depth of the cell layer containing the pigments is important to the final color. Some fruit make pigments in all their cells and others only in the external cells. Anthocyanin pigments are found in vacuoles and are greatly influenced by vacuolar pH.

Considering the large number of anthocyanin pigments identified by chemists, the major anthocyanidin of tree fruit is cyanidin. Apples and pears accumulate the 3-galactoside; sweet cherries, plums, and peaches, the 3-glucoside; and cherries, the 3 rutinoside. These are the major anthocyanin pigments, but apparently all anthocyanin-producing plants make minor pigments as well, and the number known is dependent on how extensively they have been sought. Species-specific chemistry is a significant tool in plant identification, and an entire field of specialists in chemotaxonomy exist. The chemotaxonomy of processed fruit can be employed to expose adulteration.

Chemicals other than pigments affect the final perceived color of anthocyanins. These are known as copigments and include metal ions (magnesium, iron, aluminum), hydroxycinnamoyl esters, galloyl esters, and flavone and flavonol glycosides. The three classes of pigments are only synthesized in plants. Animals cannot produce chlorophylls, carotenoids, or anthocyanins.

PHYSIOLOGY OF COLOR FORMATION IN TREE FRUIT

Because of the economic importance of fruit color, environmental factors influencing its development have been carefully researched. The primary determinant for fruit color is genetics, and by selection of crosses for desired color effect, a variety of colors have been maintained and offered to consumers.

In isolated cases, a fruit cultivar desired by consumers may not accumulate enough pigments to meet some established standard. The standard may be market acceptability or may be imposed by an overseeing body such as the U.S. Department of Agriculture Fruit Inspection Service. Commodity quality standards might relate to percent color coverage of the surface or to color intensity.

Light

Chlorophyll synthesis requires light. Chlorophyll and most pigments are in a balance between synthesis and destruction. Placing fruit in darkness will result in bleaching of the green color. The result might be white or yellow fruit depending on which pigments are present. Light exclusion is commercially practiced by placing bags over developing fruit. One desirable result is the loss of green color, making the fruit look brighter. The bags must be changed or removed to allow light for anthocyanin synthesis. Controlled atmosphere storage, even for short periods of time, will reduce chlorophyll loss, which is advantageous for green apple cultivars.

Anthocyanin synthesis is light dependent in many but not all fruit. The radiation intensity required for color development is species specific, and some fruit are not considered to require light. The light requirement is in addition to photosynthesis, and although similar in spectral response to the light reactions of phytochrome, it is distinct and high energy. The light response is not translocated from cell to cell, so maximum exposure of the fruit is the best solution. Tree training systems have been developed to maximize the quantity of light energy reaching the fruit. Reflective materials spread under the tree increase the amount of light reaching the fruit and therefore increase color. For many years it has been known that apples will color after harvest, but as initially practiced by spreading the apples under the tree for days, there was significant loss of storage quality.

Temperature

The influence of temperature on fruit color has been widely studied. Poor color of apples was the traditional problem. 'McIntosh' apples, in addition to a light requirement, color better during cool temperature conditions than during warm periods. The mechanism appears similar to "autumn coloration" of foliage, which also involves senescence and is enhanced by cool temperatures. A proposed biochemical mechanism is based on differentially temperature-dependent enzyme turnover. Tree cooling with overhead sprinklers will increase anthocyanin content in apples.

Nutrition

Excessive nitrogen fertilization generally increases the green color of tree fruit and reduces their synthesis of anthocyanins. How much is excessive is difficult to predict in any specific location. Nitrogen is detrimental in the form of traditional soil fertilizers or urea sprays.

Color attracts people to fruit, as it attracts other mammals and birds to fruit. Humans are possibly more variable or more fickle about fruit color choices because some prefer green, some red, some pink, some striped, some cheeked, some solid colored, some with red flesh, some white flesh, some pink flesh, and on and on. Some even prefer the lack of color found in russeted fruit. As more beautiful fruit colors are bred and dependably produced, more consumers will be attracted to supermarket fruit displays.

Related Topics: BREEDING AND MOLECULAR GENETICS; FRUIT MATURITY; LIGHT INTERCEPTION AND PHOTOSYNTHESIS; NUTRITIONAL VALUE OF FRUIT; SOIL MANAGEMENT AND PLANT FERTILIZATION; TREE CANOPY TEMPERATURE MANAGEMENT

SELECTED BIBLIOGRAPHY

Cooper-Driver, G. A. (2001). Contributions of Jeffrey Harborne and co-workers to the study of anthocyanins. *Phytochemistry* 56:229-236.

Lancaster, J. E. and D. K. Dougall (1992). Regulation of skin color in apples. *Critical Rev. in Plant Sci.* 10:487-502.

Singha, S., T. A. Baugher, E. C. Townsend, and M. C. D'Souza (1991). Anthocyanin distribution in 'Delicious' apples and the relationship between anthocyanin concentration and chromaticity values. *J. Amer. Soc. Hort. Sci.* 116(3): 497-499.

Wilson, M. F. and C. J. Whelan (1990). The evolution of fruit color in fleshy-fruited plants. *Am. Nat.* 136:790-809.

3

Fruit Growth Patterns

Alan N. Lakso
Martin C. Goffinet

To understand the development of fruit crops, one must first understand the relationship between the fruit and the flower(s) from which it comes. After all, the fruit can be defined as the ripened ovary of the flower, with or without other adherent floral parts, other flowers, or inflorescence structures. The ovary is that part of the female floral structure (the pistil) that contains the rudimentary seeds (ovules). Each ovule becomes a seed only after sperm transmitted from germinating pollen grains fertilizes the egg cell in the ovule. Although the term "fruit" is used in many ways, common temperate zone tree fruit develop from single flowers, even if the flowers occur in clusters. The growth patterns of fruit are in great part determined by the organization and growth potential of the floral organs and any other associated structures that contribute tissue to the mature fruit.

WHAT IS A STONE OR POME FRUIT?

In the case of cherry (typical of stone fruit, such as peach, plum, or apricot), the fruit develops only from the ovary of a single flower (Figure F3.1). No other floral or nonfloral tissue contributes to mature fruit structure. The floral organs surrounding the ovary are free from the ovary and are attached to the floral stalk, or receptacle, below the ovary. The ovary is thus "superior" to such parts. The bases of the flower's sepals, petals, and stamens form a floral cup, or hypanthium, which surrounds the central female structure, or pistil. The ripened ovary of stone fruit typically contains only one large seed, from only one of the flower's two original ovules. The mature ovary wall surrounds the seed,

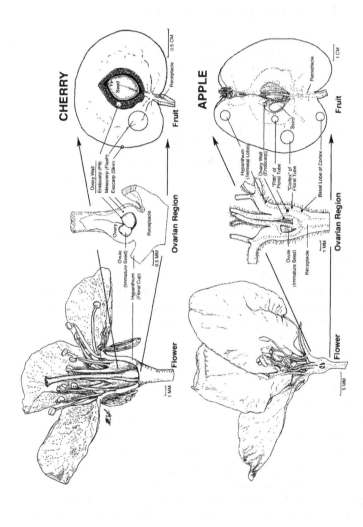

CHERRY

Flower

1 MM

Ovule
(Immature Seed)

Hypanthium
(Floral Cup)

Receptacle

0.5 MM

Ovary

Ovarian Region

Fruit

Ovary Wall
Endocarp (Pit)
Mesocarp (Flesh)
Exocarp (Skin)

Seed

Receptacle

0.5 CM

APPLE

Flower

5 MM

Ovule
(Immature Seed)

Receptacle

1 MM

Ovarian Region

Fruit

Hypanthium
(Terminal Lobe)

Ovary Wall
(Endocarp)

"Pith" of
Floral Tube

"Cortex" of
Floral Tube

Basal Lobe of Cortex

Seed

Receptacle

1 CM

FIGURE F3.1. Fate of floral tissues in development in a stone fruit (cherry) and pome fruit (apple). The wall of the superior ovary in the flower of cherry develops into the three regions of the mature fruit's pericarp, with the innermost layer becoming the lignified pit region. The receptacle and hypanthium contribute no tissue to the fruit. In apple, extracarpellary floral tissues provide most of the edible flesh that surrounds the five seed locules in the inferior ovary. These nonovarian tissues embed the ovary wall (leathery endocarp) deep within the core of the apple. Both receptacle and hypanthium regions of the flower contribute to the mature fruit flesh.

92

composes the flesh and hard pit, and is called the pericarp. The pit is the lignified inner pericarp, the endocarp. Most of the juicy fruit flesh is derived from the middle ovary wall, or mesocarp, and the skin, from the outer ovary wall, or exocarp.

In the case of apple (typical of pome fruit, such as pear and quince), the identification of the origin of fruit tissues is less obvious. The ovary region of the apple flower consists of both the ovary itself and surrounding adherent portions of the flower. Some view those surrounding tissues as portions of the flower's floral cup (hypanthium); others see those tissues as taking origin from the stem supporting the flower, the receptacle. It is from this latter interpretation that we often apply the stem terms "cortex" and "pith" to the outer and inner fleshy regions of the mature fruit. Regardless, the ovary and surrounding tissues are situated below the floral cup and the five free stigmas surmounting the ovary. Pome fruit thus have an "inferior" ovary. As the fruit grows, the apple flesh is derived to a great extent from the nonovarian tissues surrounding the embedded ovary. The ovary proper develops as the cartilaginous inner tissue (endocarp) at the fruit's core. The main flesh of the fruit is derived from the inner (pith) and outer (cortex) tissues that surround the ovary and its ovules (seeds). The hypanthium just above the inferior ovary of the flower will enlarge to varying degrees, to contribute flesh to the fruit's terminal lobes.

WHAT IS FRUIT GROWTH?

Although the general concept of growth is understood by most as an increase in size, before describing the growth patterns and how they are measured, it is useful to define what "growth" means scientifically. The strict biological concept of growth is an irreversible increase in dry weight (weight after all water is removed). Dry weight growth is important to consider in physiological studies of tree growth and development, as it relates to the energy required for growth. However, in commerce, tree fruit are normally sold on the basis of fruit diameter or fruit fresh weight, so growth measurements in diameter and fresh weight are common also.

The apparent pattern of growth of a fruit may vary depending on whether fresh weight, dry weight, or diameter is considered. Volume and weight represent all three dimensions of the fruit, while diameter represents only one dimension. So, interpretations of growth are affected by the measurement used. For example, if an apple fruit near harvest grows in diameter from 75 to 80 millimeters, that represents about a 7 percent increase in diameter, but a 20 percent increase in fresh weight growth. Consequently, growth curves of fruit size over a season will be presented in all three expressions of diameter, fresh weight, and dry weight.

Another consideration that is important to understanding fruit growth is the "growth rate," which is the amount of growth per time (day, week, etc.). If developing fruit are given no or few limitations (no competition from other fruit, healthy tree, etc.), they will grow near their maximum rates. Such growth rate patterns probably represent the inherent genetically controlled pattern of "demand" of the fruit for support from the tree. Variations from a maximum growth rate can help identify competition among organs within a tree (shoots, roots, fruit, wood) for resources that may reduce fruit growth. In practice, fruit growers try to balance the competition within trees to allow good yields and fruit quality while still maintaining good vegetative growth and annual flowering.

GROWTH BY CELL DIVISION
AND CELL EXPANSION

Fruit tissue basically grows in two ways: by producing new cells (cell division) and by having those cells expand in size (cell expansion) (Coombe, 1976). In most fruit, cell division occurs in the first several weeks after flowering and represents the first 20 to 35 percent of a fruit's growing season. Cell division occurs in many cells in the fruit simultaneously, and as the number of cells increases, there are more cells to divide. For example, two cells divide to make four cells that can divide to make eight cells, and so on. Consequently, the number of cells and the fruit weight increase at a faster and faster rate in the early season. This is called the "exponential phase" of growth.

Before cell division is completed, a transition to growth by cell expansion begins in the cells that have completed cell division. When cell division is completed, all growth is then by cell expansion, which accounts for the majority of fruit growth. Final fruit size is dependent on cell numbers and cell size as well as production of intercellular air spaces.

GROWTH PATTERNS OF DIFFERENT FRUIT

Sigmoid Growth Pattern

Different fruit grow in different ways, although there are a few general patterns of fruit growth. One common pattern is the "sigmoid" form of growth in which the fruit begins to grow slowly initially after bloom but then grows increasingly rapidly (see diameter curve in Figure F3.2). The growth rate is the greatest in midseason, followed by a slowing growth as harvest approaches. This type of growth pattern means that the growth rate in weight gain per day is low early and late in the season and greatest in midseason. Apples and pears often show a seasonal growth pattern that at first appears to be sigmoid due to a slowing of growth in cooler temperatures near harvest, although there is normally an extended linear portion of growth in midseason.

Expolinear Growth Pattern

A pattern of growth similar to the sigmoid pattern is called the "expolinear" pattern of growth (see weight curves in Figure F3.2).

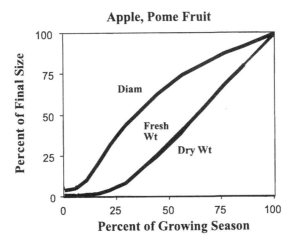

FIGURE F3.2. General seasonal pattern of diameter and fresh and dry weight growth of apples and other pome fruit as percent of the final harvest value. Low temperatures, heavy crops, or other stress may cause a late-season decline from this optimal growth pattern.

This pattern shows an early exponential curvilinear phase during cell division similar to the sigmoid pattern. However, the cell expansion phase is linear for the rest of the season. Thus, the combination of exponential and linear phases led to the name "expolinear." If apple fruit are allowed to grow with no significant limitations (i.e., with few competing fruit on a healthy tree), the fresh and dry weight growth of apples is expolinear. This pattern is also seen in Japanese or Nashi pears. Many times growth is not linear all the way to harvest (e.g., if there is heavy competition from other fruit or a limiting environment, such as cool temperatures or shorter days that cause a slowing of growth late in the season).

Double-Sigmoid Pattern

Another common pattern exemplified by stone fruit also has an initial exponential growth phase during early cell division (Figure F3.3). The fruit growth then slows significantly in midseason and completes a first sigmoid phase. A second rapid increase in diameter and fresh

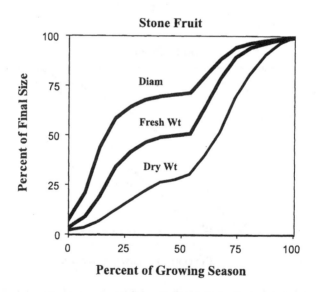

FIGURE F3.3. General seasonal pattern of diameter and fresh and dry weight growth of stone fruit as percent of final harvest value. Note that the midseason decline in growth is not as pronounced for dry weight as it is for diameter or fresh weight.

weight growth occurs next, followed by a final slowing of growth as harvest approaches. Together, the seasonal pattern is described as a "double sigmoid."

Unlike the pome fruit pattern, the seasonal pattern of growth of fruit dry weight in stone fruit is somewhat different from the pattern for fresh weight. During the slow period of fresh weight growth in midseason, dry weight growth continues. At this time, growth, called "pit hardening," is primarily in the dense tissues of the seed. The second sigmoidal growth phase is usually related to softening of the fruit and rapid accumulation of sugars. The causes and triggers of this complex pattern of growth are not known.

When considering the energy in the form of tree photosynthesis required for fruit growth over the season, it is theoretically best to measure the energy value of the fruit. However, since that is not easy, the most practical expression of growth is the daily dry weight gain, as it represents most closely the energy in most fruit that contain primarily starch and sugars. For the two main growth patterns described earlier, the seasonal pattern of dry matter growth per day is very different (Figure F3.4). The expolinear growth of pome fruit leads to a rapid increase in dry weight gain per day for the first half or so of the

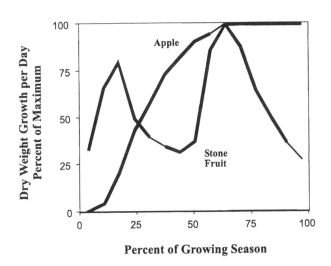

FIGURE F3.4. General dry weight growth rate per day for apples and stone fruit, representative of expolinear and double-sigmoid growth patterns. Dry matter growth gains indicate seasonal resource requirements.

season, followed by a quite constant rate. The pattern for the double-sigmoid growth of stone fruit, however, reflects the two phases of rapid growth with a slower growth rate occurring in midseason and near harvest. These curves represent general patterns for these fruit growing without significant competition and stress. The actual growth curves may vary somewhat with cultivar, and crop load or environmental stress may limit growth at times.

IMPORTANCE OF MAINTAINING FRUIT GROWTH

Studies of fruit growth and fruit set with peach and apple indicate that the fruit which remain on the plant to harvest are those which maintain their growth rates all season long. The fruit that drop, especially early in the season, are those whose growth falls to very low rates for several days. Although this has not been examined in all fruit, it appears that, in these species, adequate fruit growth is needed not only to achieve optimum final size but also to increase number of fruit remaining to harvest (and thus the yield potential).

Fruit growth can be limited by several factors. First, most fruit crops produce many more flowers than the plants can support to maturity. There is natural drop of many fruit, but without grower intervention, tree fruit will still produce an overly large crop of very small fruit. The numerous fruit compete for limited resources from the tree (probably carbohydrates or nitrogen), and consequently fruit growth is inhibited. This is especially critical during the early cell division period. If competition can be limited by reducing the number of fruit at this time, final fruit size will be improved. Because of this response, considerable research is devoted to developing procedures for thinning fruit (reduction of fruit numbers to increase size). Thinning is a common and critical practice in production of most temperate tree fruit crops.

Besides competition among fruit on the same plant, there may be many environmental limitations to fruit growth. The most obvious is temperature. Fruit growth in the early season during cell division appears to be quite sensitive to temperature, similar to many growth processes. However, once fruit begin to grow by cell expansion, fruit growth rates appear to be much less influenced by temperature. Reduced light availability, during dark, cloudy periods or by shade within a plant interior, also limits fruit growth, especially in the early season. This effect appears to result from reduced availability of carbohydrates to fruit. Low light can cause a dramatic reduction in fruit set, particularly if there is a heavy crop on the plants at the time. In addition, any resource limitation such as drought stress or nutrient deficiency can limit fruit growth, just as it limits all forms of plant growth (shoots, roots, etc.).

FRUIT SHAPE

Fruit weight or diameter is not the only important result of fruit growth. Fruit shape may change over the season of growth, giving unique forms to different fruit. The general shape of fruit is determined relatively early in fruit development (i.e., young fruit generally look very similar to mature fruit). Several fruit, however, have distinctive features. Examples are peaches, apricots, and other stone fruit that have a pronounced suture groove running the length of the fruit, and the 'Delicious' apple, with its distinct conical shape and five lobes.

Changes in fruit shape over the season are the result of differential growth of tissues in particular locations in the fruit, such as the lobes in the 'Delicious' apple (Figure F3.5). The difference in growth may be due to more or less growth in certain tissues, to a difference in the type of cells produced (e.g., smaller, more compact cells), or to the relative growth differences in one dimension versus another. For example, because the relative growth of apple flesh is somewhat greater in the radial direction than in the longitudinal direction, the shape of an apple tends to become more oblate (wider) as it grows. Many fruit develop under hormonal stimuli from seeds, so that a failure of seed development may alter fruit shape dramatically.

Temperate tree fruit exhibit several different patterns of fruit growth. Within a species, the pattern of fruit growth depends on the dimension monitored—diameter, fresh weight, or dry weight. Diameter and fresh weight are important in commerce, but dry weight is more important to the physiology of the tree.

Related Topics: CARBOHYDRATE PARTITIONING AND PLANT GROWTH; FLOWER BUD FORMATION, POLLINATION, AND FRUIT SET; LIGHT IN-

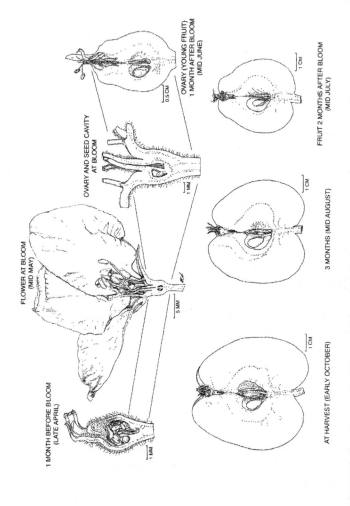

FIGURE F3.5. Development of size and form of 'Delicious' apple fruit from its inception in the flower to the fruit at harvest

TERCEPTION AND PHOTOSYNTHESIS; TEMPERATURE RELATIONS; WA-TER RELATIONS

SELECTED BIBLIOGRAPHY

Bollard, E. G. (1970). The physiology and nutrition of developing fruits. In Hulme, A. C. (ed.), *The biochemistry of fruits and their products* (pp. 387-425). London, England: Academic Press.

Coombe, B. G. (1976). The development of fleshy fruits. *Ann. Rev. Plant Physiol.* 27:508-528.

DeJong, T. M. and J. Goudriaan (1989). Modeling peach fruit growth and carbohydrate requirements: Reevaluation of the double-sigmoid growth pattern. *J. Amer. Soc. Hort. Sci.* 114:800-804.

Lakso, A. N., L. Corelli-Grappadelli, J. Barnard, and M. C. Goffinet (1995). An expolinear model of the growth pattern of the apple fruit. *J. Hort. Sci.* 70:389-394.

– 4 –

Fruit Maturity

Christopher B. Watkins

Several phases are recognized in the development of horticultural crops from initiation of growth to death of a plant or plant part. These are growth, maturation, physiological maturity, ripening, and senescence. Watada et al. (1984) define useful terminology for these developmental stages for understanding fruit maturity. "Growth" is the irreversible increase in physiological attributes (characteristics) of a developing plant or plant part. "Maturation" is the stage of development leading to the attainment of physiological or horticultural maturity. "Physiological maturity" is the stage of development at which a plant or plant part will continue ontogeny, even when detached. "Ripening" is the composite of the processes that occur from the latter stages of growth and development through the early stages of senescence that results in characteristic aesthetic and/or food quality, as evidenced by changes in composition, color, texture, and other sensory attributes. "Senescence" involves those processes which follow physiological maturity and lead to death of tissue. The developmental stages overlap; those between maturity, ripening, and senescence are particularly important in a discussion of temperate fruit maturity, as these events can be temporally close.

An additional term that is used in discussion of maturity is "horticultural or harvestable maturity." This is a relative term representing a stage of development when a plant possesses the prerequisites for utilization by consumers for a particular purpose. Thus, many commodities may be harvested when physiologically immature. Temperate fruit, however, usually are harvested when fully developed and physiologically mature. At the time of harvest, ripening may also have occurred, but additional ripening can be required to meet consumer requirements. A mature fruit can be defined as one that has reached a stage in its growth and development cycle that, after harvesting and postharvest handling (including ripening, when required), will be at least the minimum quality acceptable to the consumer (Reid, 1992). Immature fruit may not ripen to meet flavor requirements of the consumers. They also may be prone to the development of physiological disorders, for example, bitter pit and superficial scald in apples, shriveling and friction discoloration in pears, and chilling injury in stone fruit. Overmature fruit, in contrast, may have fuller flavor, but texture can be poor and storage periods restricted because of susceptibility to injury and decay. Disorders associated with overmaturity may also develop, including physiological disorders such as soft scald and watercore in the case of apples, and susceptibility to internal injuries associated with low oxygen or elevated carbon dioxide in the storage atmosphere in the case of apples and pears.

THE COMPROMISE BETWEEN QUALITY AND STORABILITY

The quality of any horticultural crop is a combination of attributes that provides value in terms of human consumption. Depending on the point in the marketing chain, however, the concepts of quality can vary. Shippers and packers are concerned with appearance and absence of defects, receivers and distributors with firmness and storage life, whereas consumers perceive quality based on appearance, nutritive value, and eating quality factors such as texture and flavor.

The dilemma for horticultural industries is that as the quality attributes associated with development of a ripe, edible fruit are increasing, the storability of the fruit is decreasing (Figure F4.1). This change is often associated with the increase in ethylene production that occurs during ripening of climacteric fruit. Each industry has to establish the appropriate compromise between increasing quality of fruit and storability. The decisions on when to harvest a fruit will depend, therefore, on market requirements and factors such as distance between the growing region and the market. Examples include the following:

1. *Cultural differences.* Consumers in Continental Europe, for example, prefer apples at a more advanced stage of ripeness than those in the United Kingdom. Asian markets have preferences

for sweeter apples such as 'Delicious' and 'Fuji' over the more acid cultivars such as 'Cox's Orange Pippin' and 'Braeburn' preferred by European markets. Asian markets also have a greater acceptance of the "disorder" watercore, which is associated with more mature fruit, while in European markets, it is considered a defect rather than a positive attribute.

2. *Storage length and transport distance.* A fruit destined for long-term controlled atmosphere storage or for transport, e.g., from the Southern to Northern Hemisphere, will have to be harvested at an earlier stage of maturity than a fruit that is harvested for immediate consumption.

3. *Consumer acceptance.* Sensory requirements of the consumer may change according to the time of year that fruit are purchased. Greater flavor, associated with tree-ripened fruit, is likely to be a premium factor in "pick your own" or gate sales during autumn. In contrast, for long-term stored fruit, earlier harvest is required because texture is more likely to be a critical acceptance factor.

DEVELOPMENT OF MATURITY INDICES

A maturity index should relate consistently from year to year to quality of the marketed product. The many physiological and biochemical changes that occur during maturation and ripening of apples, pears, peaches, nectarines, plums, and cherries have led to testing of an extensive range of potential maturity indices. These maturity indices have been based on different criteria, depending on the industry involved, and include development of correlations between maturity-related attributes, the progression of these attributes with advancing maturity, and the relationships between these attributes and edible quality and/or the occurrence of physiological disorders.

Because of the differences in maturation and ripening physiology within each fruit type, there can be wide variations in the "best" index, or set of indices, that is regarded as suitable for any given cultivar. In addition, adoption of certain maturity indices is affected by regional differences in technologies available for maturity assessment, the size and sophistication of the specific industry, and market requirements. In general, maturity indices should be simple; readily

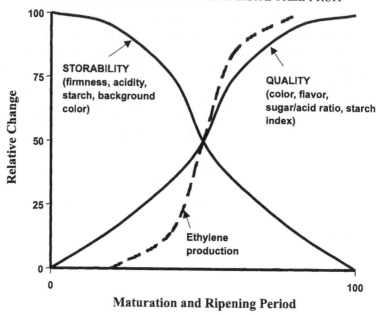

FIGURE F4.1. Schematic illustration of the increase in fruit quality during maturation and ripening, and concomittant loss of storage potential. Autocatalytic ethylene production is generally associated with these changes in climacteric fruit.

performed by growers, field staff, or industry personnel; and objective rather than subjective. Ideally, they require inexpensive equipment, but depending on the size of the industry, more expensive equipment may be used. For example, gas chromatographs for assessment of internal ethylene concentrations (IEC) are used in some apple maturity programs, where samples are consolidated across a region for evaluation in a single laboratory.

Several "maturity indices" are indicators of quality rather than maturity per se, and, in addition, harvest decisions have to be based, not only on physiological maturity, but also on market requirements. Thus, a fruit may obtain physiological maturity but, unless it meets market requirements, such as blush and background color, will not be acceptable in many markets. The term "harvest indices" is more accurate for the factors used in making harvest decisions. Of the harvest indices available for temperate fruit, several of the more commonly used ones are discussed here:

1. *The production of ethylene,* an important plant hormone, is often associated with initiation of ripening and, therefore, is sometimes used as a major determinant in harvest decisions, especially for apples. However, the importance of ethylene in making harvest decisions is not straightforward; relationships between ethylene production and optimum harvest dates can be poor, and the timing, or presence, of increased ethylene production is affected by cultivar. Moreover, within a cultivar, ethylene production is greatly affected by factors such as growing region, orchard within a region, cultivar strain, growing season conditions, and nutrition. Ethylene production may be a better indicator of when to complete the harvest, especially in cultivars where autocatalytic ethylene production precedes preharvest drop.

2. *The starch test,* in which the hydrolysis of starch to sugars as fruit ripen is estimated by staining starch with iodine solution, has become popular for assessment of apple fruit maturity. The resulting patterns, which reflect the extent of starch hydrolysis, are rated numerically using starch charts, either specific to cultivar or generic (Figure F4.2). Optimum starch indices are available for many cultivars, and because the change of indices is linear, the test can be used to predict optimum harvest dates.

3. *Flesh firmness* has been used as a maturity index, but it is affected by many preharvest factors, including season, orchard location, nutrition, and exposure to sunlight, that are independent of fruit maturity. It is the primary method for assessing maturity of pears. For other fruit, it is an important indicator of internal quality and can provide information that is important to fruit performance in storage. It can directly affect consumer satisfaction with many fruit. For apple, firmness is used as a quality criterion by wholesalers, especially in England.

4. *The soluble solids concentration* (SSC) of fruit generally increases as fruit mature and ripen, either directly by import of sugars or by the conversion of starch to sugars. It is also a quality index, rather than a maturity index, being affected by many preharvest factors, and concentrations do not necessarily reflect fruit maturity. As with firmness, SSC is increasingly being used as a quality criterion by wholesalers.

5. *Titratable acidity* (TA) primarily estimates the amount of the predominant acid, usually malic in most temperate fruit. TA decreases during maturation and ripening, but optimum values vary by cultivar and season.
6. *The background, or ground, color change* from green to yellow reflects the loss of chlorophyll. Preharvest factors, especially those which affect nitrogen content, can markedly influence chlorophyll concentrations, independent of maturity changes.
7. *Full-bloom dates and days after full bloom,* with and without incorporation of temperature records, have been established, but usefulness varies greatly by cultivar and growing region. Calendar dates alone have limited value in regions where temperature variations result in wide differences in bloom dates, but in more consistent growing regions, days from full bloom can be the most reliable harvest index for some cultivars.

Even when certain maturity indices are considered as imperfect harvest indicators, they may be useful in combination. For example, Crisosto (1994) reports that flesh firmness in combination with background color is an excellent indicator of maximum peach maturity.

MATURITY OR HARVEST INDICES
FOR SPECIFIC FRUIT

The following summary of harvest indices should be considered as an overview for each fruit type. Advice about appropriate maturity indices should be obtained from local university or extension personnel within a growing region or industry.

Apples

The most commonly used maturity indices are ethylene production or IEC, the starch test, flesh firmness, SSC, TA, background color, calendar date, days from full bloom and temperature records, and heat accumulation (Watkins, 2002). Other indices include fruit size, sliding scales of firmness and SSC, ratios of firmness/soluble solids multiplied by starch (the Streif index), the "T-stage," flesh color, seed color, loss of bitter flavor, appearance of watercore, and separation force.

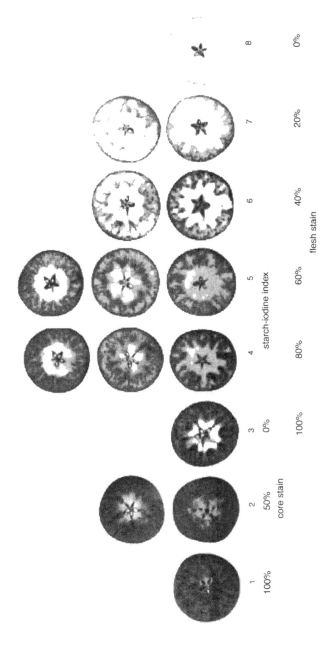

FIGURE F4.2. Generic starch-iodine chart (*Source:* Modified from Blanpied and Silsby, 1992.)

Pears

The pressure test (firmness) has proven to be the most reliable and seasonally consistent method for determining harvest maturity of all pear cultivars (Hansen and Mellenthin, 1979). Each cultivar has a specific firmness range for optimal harvest. Other indices include ground color, development of a smooth, waxy skin (especially for 'D'Anjou'), a moist rather than dry cut surface on cross-sectioned fruit, days from full bloom and temperature records, heat accumulation, the starch index, size, and optical density of fruit. SSC is unreliable as a maturity index, but a minimum concentration of 10 percent is required for best quality and prevention of freezing during storage (Hansen and Mellenthin, 1979). Fruit are usually harvested preclimacteric, and, therefore, ethylene production is not a useful maturity index for pears.

Peaches, Nectarines, and Plums

Maturity indices include fruit size and shape, ethylene production, respiration rate, firmness (cheek and blossom end), SSC, TA, SSC/TA ratio, and background color (on cheeks and blossom end). Other investigated indices include near-infrared light, magnetic resonance, light transmittance, delayed light emission, and microwave permittivity. A study of several maturity indices in plums demonstrates that none are reliable and applicable to all cultivars (Abdi et al., 1997).

Cherries

Fruit color is the most commonly used indicator of ripeness, that of black sweet cherries progressing from straw color to very light red, followed by red, which darkens to mahogany (Looney, Webster, and Kupferman, 1996). Yellow sweet cherries develop yellow flesh and skins as one of the first signs of maturity, and some cultivars are harvested when a red blush develops on the cheek. Color of black cherries is judged commercially using color comparators or cards and must be shiny, not dull, in appearance. Other indices include size, firmness, and flavor (SSC and acidity). Fruit retention strength is used as the main maturity guide for cherries that are harvested mechanically.

MATURITY PROGRAMS

Fruit-growing regions throughout the world vary in the type and extent of maturity programs. For apples, IEC measurements and the starch index have become the most widely used maturity indices, although for some bicolored apples, e.g., 'Gala', 'Braeburn', and 'Fuji', background color is considered an important harvest index. In some cases, state regulations have been established to set minimum harvest maturities, e.g., the starch index for 'Granny Smith' in California. Currently in Washington State, individual packinghouses conduct their own maturity programs in line with their marketing strategies. In Michigan and New York, a wide range of maturity and quality indices is collected, and the optimum harvest period, sometimes called the harvest window, is established each year for major cultivars (Beaudry, Schwallier, and Lennington, 1993; Blanpied and Silsby, 1992). The Streif index is used in some parts of Europe.

Programs for stone fruit are typically less formalized. There are no established maturity programs for cherries in Washington, for example, with harvest decisions being based on fruit color, market pressure, and predicted weather. Fruit bound for Asia are harvested slightly less mature than fruit sold within the United States. For peaches, nectarines, and plums, individual companies in growing regions such as South Africa, California, Chile, and Argentina accumulate data relating to harvest maturity and subsequent eating quality, arrival condition in markets, and market life.

Irrespective of the crop involved, fruit maturity is a critical factor in consumer satisfaction and impacts the effect of the many abuses that can occur during subsequent handling operations. While certain maturity indices are more important than others in establishing the correct time to harvest fruit, the maturation and ripening processes involve many simultaneous biochemical and physiological changes. The strength of any maturity program probably lies, not in reliance on absolute maturity indices, but in discussion with industry personnel on changes in maturity and quality that are occurring over the harvest period. Commercial factors such as color, susceptibility to bruising with progressing maturity, and issues such as weather patterns that will affect harvest cannot be ignored. In this way, full participation of growers, storage operators, shippers, and other industry personnel can ensure that fruit of appropriate quality are received in the market-place.

Related Topics: HARVEST; PACKING; PHYSIOLOGICAL DISORDERS; POSTHARVEST FRUIT PHYSIOLOGY; PROCESSING; STORING AND HANDLING FRUIT

SELECTED BIBLIOGRAPHY

Abdi, N., P. Holford, W. B. McGlasson, and Y. Mizrahi (1997). Ripening behavior and responses to propylene in four cultivars of Japanese type plums. *Postharvest Biol. Technol.* 12:21-34.

Beaudry, R., P. Schwallier, and M. Lennington (1993). Apple maturity prediction: An extension tool to aid fruit storage decisions. *HortTechnol.* 3:233-239.

Blanpied, G. D. and K. J. Silsby (1992). *Predicting harvest date windows for apple,* Info. bull. 221. Ithaca, NY: Cornell Coop. Ext. Serv.

Crisosto, C. H. (1994). Stone fruit maturity: A descriptive review. *Postharvest News Info.* 5:65N-68N.

Hansen, E. and W. M. Mellenthin (1979). *Commercial handling and storage practices for winter pears,* Agric. exper. sta. special report 550. Corvallis, Oregon: Oregon State Univ.

Looney, N. E., A. D. Webster, and E. M. Kupferman (1996). Harvesting and handling sweet cherries for the fresh market. In Webster, A. D. and N. E. Looney (eds.), *Cherries: Crop physiology, production and uses* (pp. 411-441). Oxon, UK: CAB International.

Reid, M. S. (1992). Maturation and maturity indices. In Kader, A. A. (ed.), *Postharvest technology of horticultural crops,* Bull. 3311 (pp. 21-30). Oakland, CA: Univ. of California.

Watada, A. E., R. C. Herner, A. A. Kader, R. J. Romani, and G. L. Staby (1984). Terminology for the description of developmental stages of horticultural crops. *HortScience* 19:20-21.

Watkins, C. B. (2002). Principles and practices of postharvest handling and stress. In Ferree, D. C. and I. J. Warrington (eds.), *Apples: Crop physiology, production and uses.* Oxon, UK: CAB International. In press.

1. GEOGRAPHIC CONSIDERATIONS

Geographic Considerations

Suman Singha

As their name implies, temperate tree fruit grow primarily in the temperate zone. However, even in this region, geography is often an important consideration in determining the success or failure of an enterprise. Whether these crops can be successfully raised in regions outside the temperate zone will also be influenced by geographic considerations, as these will have a significant impact on the climate.

LATITUDE

The temperate zone extends from approximately 35 to 60 degrees north and south of the equator, and the region has four distinct seasons—winter, spring, summer, and fall. Temperate fruit species have an endodormancy, or rest requirement, that can be satisfied under the climatic conditions of only this region, and, consequently, this factor limits the growth of these species to this zone. Dormancy is overcome by exposing the plants to a chilling period (at 4 to 7°C) in winter. The length of exposure needed to overcome endodormancy varies with species (and even cultivars within a species) but on average ranges from 400 to 1,200 hours for peaches and 800 to 1,500 hours for apples and pears. Although the winter hardiness of the different temperate species (and cultivars) varies, most cannot withstand the extreme winter cold encountered in areas beyond the latitude of the temperate zone. This is another factor limiting their growth to this region.

Although the temperature of a location is primarily a function of latitude, it is also strongly influenced by altitude. The temperature drops approximately 3.5°C for every 500 meters increase in altitude.

As a result of this, the peak of Mt. Kilimanjaro in Tanzania, which lies three degrees south of the equator at a height of 5,895 meters, has year-round snow cover, and regions in the Himalayan mountain range are major producers of temperate fruit crops.

ELEVATION

The elevation of an orchard site compared to its surroundings can be an important factor, especially where late spring frosts are a problem. Cold air settles in low-lying areas, and thus locations that are slightly higher than the surroundings have good air drainage and will be less likely than low-lying areas to suffer frost damage. Low-lying areas are prone to greater winter injury for the same reason.

WATER BODIES

Large water bodies, such as lakes, have a significant ameliorating influence on the local climate. In winter, the water serves as a heat source and keeps the surrounding areas slightly warmer than areas outside its influence. In spring, the water serves as a heat sink and keeps the surrounding areas colder. This has a significant impact in delaying bloom of fruit trees and thereby reducing potential damage from spring frosts. This is a major reason why, for example, the fruit industries in New York and Ontario are located around Lake Erie and Lake Ontario.

ASPECT

The aspect can influence the local climate. Southern-facing slopes receive more sunlight and consequently tend to be slightly warmer than locations with a more northern exposure. This can result in slightly earlier blooming in spring and potentially a higher probability of sunburn later in the season.

RAINFALL

Rainfall can become a limiting factor if supplemental irrigation is not available. Most temperate zone areas receive precipitation in the form of snowfall, and thus soil moisture is adequate in the spring. Rains in the spring in conjunction with the increasing temperatures can be problematic from the standpoint of the spread of diseases, including apple scab, during this period. Lack of rains (or irrigation) later in the season can negatively impact fruit enlargement and reduce fruit yield.

Selecting a proper site is critically important to the success or failure of an orchard. Many an operation, for instance, has failed or been rendered unprofitable because of repeated spring frost damage. Orchards are long-term investments, and site-related problems can be devastating.

Related Topics: CULTIVAR SELECTION; DISEASES; DORMANCY AND ACCLIMATION; ORCHARD PLANNING AND SITE PREPARATION; SPRING FROST CONTROL; TEMPERATURE RELATIONS

SELECTED BIBLIOGRAPHY

Childers, Norman F., Justin R. Morris, and G. Steven Sibbett (1995). *Modern fruit science.* Gainesville, FL: Horticultural Publications.
Westwood, Melvin N. (1993). *Temperate-zone pomology: Physiology and culture.* Portland, OR: Timber Press.

1. HARVEST
2. HIGH-DENSITY ORCHARDS

Harvest

Stephen S. Miller

The harvest of tree fruit is a labor-intensive task. Labor availability for harvest, once considered plentiful, now constitutes an important limiting factor for many growers in Europe, the United States, and other fruit-producing countries. Harvesting requires special attention to ensure that fruit are picked at the proper stage of maturity with minimal damage. Because most tree fruit crops are still harvested by hand, it is a major production cost. Since market destination generally dictates the method of harvest, fruit intended for the fresh market are usually hand harvested, while fruit designated for processing may be mechanically harvested.

HAND HARVESTING

Hand harvesting the highest-quality fruit requires special knowledge about the crop and the necessary harvest equipment. Most tree fruit crops, even those grown on dwarfing rootstocks, require some use of ladders for harvesting (Figure H1.1). Stepladders are generally used for smaller trees (up to 3.5 meters height), while straight ladders are used for taller trees (14.5 meters or above), although in some areas, such as Washington State, stepladders are used even for the taller trees. Traditionally, these ladders were constructed of a lightweight wood such as basswood, but more recently, aluminum has replaced wood. Aluminum ladders have a longer life and are less subject to breakage under the weight of a picker carrying a container of fruit. Some growers use only stepladders to avoid knocking fruit from the tree, which often occurs when a straight ladder is inserted into the tree canopy. Orchard stepladders (sometimes called tripod ladders) are

FIGURE H1.1. Various styles of aluminum stepladders used to hand harvest tree fruit crops

designed with three legs to provide stability on uneven ground. Both stepladders and straight ladders are built with a wide base that narrows at the top. This offers stability while minimizing resistance when inserting the ladder into the canopy. Straight ladders are carried in an upright position as pickers move around the tree and from tree to tree. Skill is required to balance the ladder in an upright position, and pickers should be instructed in the proper and safe use of ladders at the beginning of harvest.

Lightweight canvas, plastic, or sheet metal containers are used to collect fruit crops such as apple, peach, and pear as they are harvested from the tree. Containers are fitted with heavy cloth straps worn over the picker's shoulders to support the harvested fruit, thus freeing hands to climb the ladder and to harvest fruit. Over the years, most of the flexible canvas picking "bags" have been replaced by rigid picking buckets (Figure H1.2) that may be fitted with soft, padded linings. These picking containers help reduce damage to the fruit as it is being harvested and transported from the tree to the bulk collection container. Sweet cherries are hand harvested in flats, trays, or buckets (typically about 7.5-liter containers). Tart (sour) cherries for a local fresh market may also be harvested by hand, but most sour cherries are mechanically harvested.

Bruising is the primary damage associated with hand-harvested fruit. Special care must be taken in removing fruit from the tree, placing it in the picking bucket, and in emptying fruit into the bulk container. An apple or pear is harvested by grasping the fruit with the fingers while it rests in the palm of the hand, lifting it upward, and

FIGURE H1.2. A rigid canvas picking bucket commonly used to harvest apples, pears, and peaches

twisting slightly to separate the fruit stem from the spur. If an apple is harvested by pulling straight down, the stem will often be removed, and sometimes the spur will also be detached from the tree. Spur damage or detachment reduces future crops and may lead to disease, especially in stone fruit. Pickers can also cause puncture damage to fruit skin with their fingernails. Wearing gloves is recommended to alleviate this problem.

Because peaches have short stems, they may be harvested by pulling straight down or away, particularly when attached near the base of large shoots. If a peach growing near the base of a large shoot is harvested by twisting, the skin near the stem end will often be broken. Peaches attached toward the apex of thin shoots may be harvested by pulling while giving a slight twisting motion. Peaches harvested at a "tree-ripe" stage of maturity are more subject to finger bruise damage than are fruit harvested at the firm-ripe stage.

Sweet cherries have traditionally been harvested with their stems (pedicels) attached. Since the picker is not handling the fruit directly but is grasping the stem, the opportunity for fruit damage is reduced.

Stemmed cherries are visually attractive and thought to have less spoilage than stemless cherries, since the flesh is not torn at the stem and fruit junction. In the eastern United States, sweet cherries that are sold directly to the retail market are sometimes harvested without the stems attached. Stemless sweet cherries should be harvested fully mature, since the fruit must readily detach from the stem with light finger pressure.

Harvested fruit are generally emptied into large bulk containers in the field for transport to the packinghouse or the processing plant. These "bulk bins" come in various sizes and are constructed of several materials, including wood, plastic, and steel. Plastic bins are lighter, cool more quickly in storage, and do not harbor disease, which is a problem with wooden bins. When used properly, plastic bulk bins last longer than wooden bins. Most bins have slots in the sides and bottoms to allow for movement of air and water; however, solid bins are used for tart cherries since they are transported in water. To minimize bruise damage to soft fruit, shallow bins should be used or large bins should not be filled to capacity.

Harvest employees are either paid by the "piece rate" or on an hourly basis to hand harvest fruit. Piece rate payment is generally based on the size of a container. For apples, the common measure is a bushel (19 kilograms) or a bulk bin (typically a 342- to 475-kilogram container). Cherry pickers may be paid by the pail or the flat harvested. Payment by piece rate often leads to more fruit damage, since this method encourages pickers to harvest as many containers as possible within the work day. When pickers are paid an hourly rate, speed is not a priority and tonnage is sacrificed for quality.

HARVEST AIDS

Time-motion studies have shown that 30 to 50 percent of the time required to harvest tree fruit crops is spent climbing and positioning the ladder and placing harvested fruit in the bulk container. Harvesting aids can increase efficiency by 15 to 20 percent or more, depending on tree design, orchard terrain, and uniformity of fruit load. Some of the earliest harvesting aids were designed to position a single picker in the canopy of a conventional large freestanding fruit tree. The cost of these harvest aids and the problems associated with moving the harvested fruit from the picker to the bulk container have limited their use within the industry.

Most harvesting aids are designed for high-density orchards trained to a hedgerow or similar continuous canopy system. For this reason, they have been more common in European orchards than in the United States, where many orchards still consist of single-tree units planted at low or medium densities. A typical harvest aid for high-density plantings is designed to position two to six pickers at various levels in the tree canopy. Pickers work as a team as the aid moves down the tree rows. A unit consists of a towed or self-propelled base with platforms constructed at several levels, which are fitted with telescoping catwalks that can be extended into the tree. Systems to convey fruit to a bulk container and for handling the bulk container are built into the harvest aid. Recent designs have incorporated computerized self-steering mechanisms and improved conveyor systems for handling fruit between the picker and the bulk container (Figure H1.3).

MECHANICAL HARVESTING

Widespread interest in mechanical harvesting in the United States began in the early 1950s with the gradual decline in a readily available, qualified seasonal labor pool to hand harvest tree fruit. The majority of tree fruit mechanically harvested are destined for the processing market, especially tart cherries. Success in adapting mechanical harvesting

FIGURE H1.3. Mechanical harvest aids: (left) a one-person motorized tower and positioning aid, with forks for carrying a bulk container; (right) a self-propelled computerized harvesting aid with two pickers for inclined canopy trees

techniques to apple and other deciduous tree fruit crops has, however, been somewhat limited, especially for fresh-market-quality fruit. A major obstacle to mechanical harvesting has been excessive fruit damage. Since the 1980s, significant effort has been made to develop mechanical harvesters for fresh-market apples and to some extent peaches. However, with peaches, a lack of uniform maturity is a major problem. Recently, work has been directed toward harvesting fresh-market-quality sweet cherries. While progress has been made, a successful mechanical harvester for fresh-market apple, peach, and sweet cherry has not been developed.

Most commercial mechanical harvesters for tree crops operate on the shake-and-catch principle. A large clamp is attached to the tree trunk or to an individual limb. Fruit are detached, collected on a padded surface, and conveyed to a bulk container. The catching surface incorporates various deflectors, rollers, and positioning devices to reduce the impact of falling fruit and the damage from fruit-to-fruit contact. Various rotary inertial or recoil impacting devices built into the clamping head affect detachment. Most rotary shakers use a multidirectional action in the shaker device; impactors are unidirectional. Many commercial mechanical harvesters consist of two self-propelled units or halves, one on each side of the tree (Figure H1.4). One unit contains the shaking mechanism, a collecting surface, a conveyor system, and the bulk container. The other unit consists mainly of a collecting surface but may also have a conveying system. Some mechanical harvesters are single units with wraparound frames. These units resemble inverted umbrellas. In the United States, most tart cherries are mechanically harvested with two-half, inclined-plane, shake-and-catch harvesters. Canning peaches, and, to a lesser extent, apples for processing are also harvested with this type of equipment. The use of shake-and-catch mechanical harvesters for processing apples has declined since the 1990s, as processors have demanded higher-quality fruit for canning purposes. The shake-and-catch harvester principle has also been designed into over-the-row continuous moving units that can straddle small-stature trees. These units are more efficient, but fruit damage levels are similar to those obtained with the larger, two-half harvesters.

Skill is required to operate mechanical harvesters to avoid damage to trees as well as the harvested fruit. Trunk- or limb-shaking units place tremendous pressure on a tree's bark and cambium during the

FIGURE H1.4. A shake-and-catch mechanical harvester being used to harvest cling peaches. This unit employs a trunk shaker to detach fruit from the tree.

clamping and shaking operation. Irrigation should be halted several days prior to mechanical harvesting to reduce bark slipping. Clamping pads in the shaker heads must be lubricated periodically to avoid removing bark from a tree. Minimizing the duration of the shaking action is also important.

Compatible tree structures are considered necessary for successful mechanical harvesting. Smaller trees are more easily adapted to mechanical harvesters than large trees, but studies have shown that fruit damage may still be unacceptable with current state-of-the-art mechanical harvesters. Tree form can be adjusted to enhance fruit detachment and reduce fruit damage. With apple, recent work has concentrated on inclined trellis canopy forms that provide a more uniform, open canopy with easier access for the shaker or impacting mechanism and a clear path for fruit to the catching surface. When fruit are borne on long, slender branches, much of the energy applied to the tree is lost and fruit detachment is difficult. Pruning and training methods should encourage compact, stiff growth for easier fruit detachment. Spur-type apple trees are considered ideal for adapting to mechanical harvesting, since fruit are borne on short, stiff spurs.

Robotics has been incorporated into the latest experimental mechanical harvesters. In one such unit, television cameras mounted on the harvester feed information to an onboard computer, which then directs a robotic arm to the location of an individual fruit. Once the fruit is identified, a suction cup grasps the fruit; the arm rotates and removes the fruit, placing it into a conveyor. Another robotic bulk harvesting system designed for inclined trellis canopies uses sensors and intelligent adaptive technology along with a limb impactor (rapid displacement actuator) to locate and detach fruit and position the catching surface (Peterson et al., 1999). Although these units are still in developmental stages, the potential for mechanical harvesting of fresh-market-quality tree fruit appears promising.

Harvesting is the climax of the growing operation and a labor-intensive step in bringing a tree fruit crop to market. Most tree fruit are still hand harvested to ensure the highest possible quality, but mechanical harvesters have been developed and are used for processing fruit, such as tart cherries and canning peaches. Research is ongoing to develop mechanical harvesters capable of harvesting fruit equal to the hand picker.

Related Topics: FRUIT MATURITY; HIGH-DENSITY ORCHARDS; MARKETING; PROCESSING; TRAINING SYSTEMS

SELECTED BIBLIOGRAPHY

Brown, G. K. and G. Kollar (1996). Harvesting and handling sour and sweet cherries for processing. In Webster, A. D. and N. E. Looney (eds.), *Cherries: Crop physiology, production and uses* (pp. 443-469). Oxon, UK: CAB International.

Morrow, C. T. (1969). Research and development on harvesting aids for standard-size trees in Pennsylvania. In Light, R. G. (ed.), *Proceedings New England apple harvesting and storage symposium, 1968,* Pub. 35 (pp. 42-50). Amherst, MA: Univ. Massachusetts Coop. Ext. Serv.

Peterson, D. L. (1992). Harvest mechanization for deciduous tree fruits and brambles. *HortTechnol.* 2:85-88.

Peterson, D. L., B. S. Bennedsen, W. C. Anger, and S. D. Wolford (1999). A systems approach to robotic bulk harvesting of apples. *Trans. ASAE* 42:871-876.

Peterson, D. L., S. S. Miller, and J. D. Whitney (1994). Harvesting semidwarf free-standing apple trees with an over-the-row mechanical harvester. *J. Amer. Soc. Hort. Sci.* 119:1114-1120.

Robinson, T. L., W. F. Millier, J. A. Throop, S. G. Carpenter, and A. N. Lakso. (1990). Mechanical harvestability of Y-shaped and pyramid-shaped 'Empire' and 'Delicious' apple trees. *J. Amer. Soc. Hort. Sci.* 115:368-374.

Sarig, Y. (1993). Robotics of fruit harvesting: A state-of-the-art review. *J. Agric. Engin. Res.* 54:265-280.

Tukey, L. D. (1971). Mold the tree to the machine. *Amer. Fruit Grower* 91:11-13, 26.

2

High-Density Orchards

Suman Singha

Orchards have undergone significant changes due to the wide-spread usage of dwarfing rootstocks. Older blocks of 250 trees per hectare have been replaced by plantings of 600 to 2,500 trees per hectare. Ultra-high-density systems, with more than 5,000 trees per hectare, also are under test in some regions. The transition to high-density, or intensive, orchard systems is a systemic change that encompasses far more than simply an increase in tree density, however. It requires a re-evaluation of all orchard practices and operations, and higher managerial skills. The failure to assess a high-density orchard from a total systems approach can lead to erroneous decisions and a failure of the enterprise.

The primary reasons for the adoption of high-density orchard systems have been earlier cropping and higher yields, which translate to higher production efficiency, better utilization of land, and a higher return on investment. Trees in high-density orchards may be free-standing, staked, or supported by a trellis. This is a function of the training system, the tree species, the rootstock and cultivar selected, and the goals of the enterprise. Intensive orchards require a greater outlay of capital, labor, and managerial skills, especially during establishment. The need for greater investment is a function of the larger number of trees and tree supports and will be especially significant if a wire-supported training system is proposed.

LIGHT ENVIRONMENT

One of the primary advantages of a high-density system is increased light interception by the tree canopy. Light distribution within the canopy is a determining factor in flower bud development and vigor. Portions of the canopy that receive less than a third of the ambient levels have a significant reduction in flower bud formation, spur vigor, and spurs that produce flowers. Furthermore, given the influence of light on fruit quality, the few fruit present in the shaded interior of the canopy of standard trees have poor color and lower marketability. High-density systems with smaller tree canopies and better light interception overcome the limitations encountered with standard-size trees.

Light interception in intensive orchards is influenced by both the training system and the foliage density. Most trellised systems have canopies in an almost two-dimensional configuration, and those with vertical or inclined orientations have better exposure to light than training systems with more globular tree forms.

PRODUCTION EFFICIENCY AND PACKOUT

High-density orchards come into bearing early and have a greater yield per unit land area than conventional plantings. Individual trees have higher production efficiencies and reduced amounts of non-productive wood. Cumulative yield per hectare is related to tree density during the initial life of the orchard. Thus, increasing planting density can, up to a point, increase total yield. However, besides cumulative yield, it is important to consider yield per tree and yield efficiency (yield per unit of trunk cross-sectional area) to more accurately compare different high-density systems. Although a slender spindle planting will have a greater cumulative yield during the initial years than a low trellis hedgerow with half as many trees per hectare, yield per tree and yield efficiency tend to be higher in the latter.

High-density systems have better fruit quality than low-density systems, and this is best expressed in terms of fresh fruit packout, which is a true measure of marketability and thus profitability. Improved tree canopy light interception results in a higher percentage of extra fancy and fancy grade fruit (Figure H2.1). The downgrading of fruit because of disease and insect damage is also reduced due to an improved environment within the tree and better spray coverage.

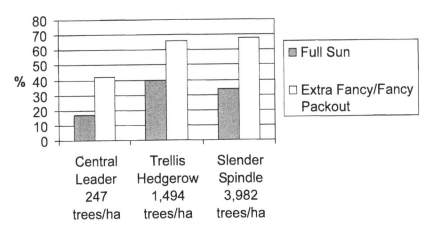

FIGURE H2.1. Tree density and training system effects on 'Golden Delicious' light transmission and packout (*Source:* Modified from Baugher et al., 1996.)

MANAGEMENT

High-density orchards allow for more efficient utilization of labor, as the trees are smaller, and it is easier to conduct routine orchard operations. However, the initial establishment of the orchard requires a greater investment in labor for tree planting and installing posts or trellises to support the trees. Also, the initial training is generally detailed as regards limb orientation and arrangement and can be labor-intensive. Overall, intensive systems require higher managerial skills than low-density orchards.

Once established, high-density orchards with smaller trees require less labor per unit of fruit produced than low-density orchards. Smaller trees and readily accessible canopies are easier to harvest, and the need for using and transporting ladders is minimized or eliminated. The ability to harvest most of the fruit from ground level is also valuable in pick-your-own operations where the absence of ladders reduces concerns of liability. Pruning is less labor-intensive in many systems, and the trees are easier to manage, provided plant growth is regulated by early and regular cropping.

Pest control in high-density orchards is facilitated because tree canopies are smaller and in many systems (especially trellised ones) are not very deep. This allows enhanced spray penetration into the canopy and reduces the need for large orchard spray equipment. Studies on spray deposits have shown that the coverage of spray materials on the leaf surfaces of trellised and nontrellised high-density trees is better than on standard trees. This, however, varies with the training system. For example, horizontal canopies have reduced spray penetration compared to vertical ones and consequently may have higher insect and disease damage.

Before selecting a specific intensive orchard system, a grower should develop clear goals, prepare a marketing plan, and conduct a cost-benefit analysis. The initial investment will vary depending on planting density, the need for tree supports, and establishment costs associated with the training system selected. Systems that produce early, have a good return on investment, and are easier to manage are essential to ensuring a profitable enterprise.

Related Topics: DWARFING; LIGHT INTERCEPTION AND PHOTOSYNTHESIS; ROOTSTOCK SELECTION; TRAINING AND PRUNING PRINCIPLES; TRAINING SYSTEMS

SELECTED BIBLIOGRAPHY

Barritt, B. H. (1992). *Intensive orchard management.* Yakima, WA: Good Fruit Grower.

Baugher T. A., H. W. Hogmire, A. R. Biggs, G. W. Lightner, S. I. Walter, D. W. Leach, and T. Winfield (1996). Packout audits of apples from five orchard management systems. *HortTechnol.* 6:34-41.

Byers, R. E., H. W. Hogmire, D. C. Ferree, F. R. Hall, and S. J. Donahue (1989). Spray chemical deposits in high-density and trellis apple orchards. *HortScience* 24:918.

Corelli-Grappadelli, L. (2001). Peach training systems in Italy. *Penn. Fruit News* 81:89-95.

1. INSECTS AND MITES
2. IRRIGATION

Insects and Mites

Tracy C. Leskey

The most important pests of stone and pome fruit are persistent and cause serious economic damage annually if they are not controlled. They belong to two classes found within the phylum Arthropoda, Insecta (insects) and Arachnida (mites). These pests can be divided into two categories, direct and indirect pests. Direct pests attack fruit and fruit buds, causing immediate injury. In some cases, damage is cosmetic, not affecting nutritional value or flavor, but diminishing aesthetic quality for marketing purposes. Indirect pests attack foliage, roots, limbs, or other woody tissues, leading to problems such as reduced tree vigor, fruit size, and/or quality and susceptibility to opportunistic secondary infections. Each growing region is prone to injury from a unique complex of pests. The pest species and groups described here are those considered to be of greatest concern on a global scale.

DIRECT PESTS

Cherry Fruit Flies

Rhagoletis species (Family Tephritidae) are pests of both sweet and tart cherries in Europe and North America. Adult flies are approximately 5 millimeters in length with distinctive black markings on wings, used to distinguish among species. Eggs are white, oval-shaped, approximately 1 millimeter long, and are inserted beneath the skin of ripening fruit by females. Eggs hatch into white maggots that feed within the fruit and drop to the ground to pupate. They construct golden brown puparia in which they will overwinter and emerge as adults the following summer. Yellow sticky traps hung in foliage of cherry trees are used to monitor adult flies; adding an ammonium-based bait can increase trap effectiveness. Organophosphate-based insecticide sprays are considered to be the best method of control, although treatment must be made within roughly one week after detection of adults in traps.

Codling Moth

Codling moth, *Cydia pomonella* (Linnaeus), is a pest of apples and pears and has been found in plums, peaches, and cherries. Present throughout Europe, Asia, and North America, this small moth belonging to the family Tortricidae is approximately 10 millimeters in length with a wingspan of 12 millimeters. Adults can be distinguished by forewing coloration and patterns; forewings are gray with bronzed tips and crossed with alternating white and gray bands. Flattened, oval-shaped eggs are laid singly on fruit and foliage. Cream to pinkish larvae hatch from eggs and enter through the calyx end or side of the fruit, feeding internally for several weeks and destroying the fruit. Mature larvae crawl from fruit on the tree or from fallen fruit to pupation sites beneath the bark of trunks and limbs. Pupae are brown and found within silken cocoons. There are two generations throughout most of the world, with mature larvae overwintering within cocoons beneath bark, leaf litter, or other sheltered areas. Adult moths emerge the following spring, beginning during full bloom to late petal fall in apple. Monitoring with pheromone traps or with pheromone traps supported by a degree-day model to predict egg hatch are effective methods for timing of insecticide application. Traditionally, organophosphate-based insecticide sprays have been used for codling moth control, although alternatives do exist, such as insect growth regulators and mating disruption. Mating disruption prevents adult males from locating females; dispensers releasing a synthetic version of the female-produced sex pheromone are attached to many trees within the orchard and serve to confuse males as they attempt to locate receptive females in order to mate. This method of control can be used in orchards that are at least 2 hectares in size. No male moths should be captured in pheromone monitoring traps if mating disruption is working well; however, mating disruption effectiveness must be determined by assessing injury.

Fruit Piercing Moth

Fruit piercing moth, *Eudocima* species, belonging to the family Noctuidae, is found throughout the Pacific Basin, Asia, India, and Africa. This moth attacks both stone and pome fruit, as well as most tree and vine crops, including citrus and many vegetables. Adults are large moths, approximately 50 millimeters in length with a wingspan of 100 millimeters. Forewings are mottled brown, green, gray, and white, while hindwings are bright orange with black borders and an oval- or kidney-shaped mark. Eggs are laid on the undersides of leaves and bark of host plants belonging to the Fabaceae and Menispermaceae families. Larvae feed on foliage of plants and pupate within a cocoon spun between leaves. These insects are unusual in that the adult is the damaging life stage; adults use a strong proboscis, approximately 25 millimeters in length, to penetrate both unripe and ripe fruit to feed on juices at night. Fruit damaged by adults degrade rapidly and provide sites for secondary rot infections. Given that larvae of these moth species feed predominantly on plants outside orchards, traditional insecticidal control is difficult. Control of adult moths is especially challenging due to their large size and limited contact with treated plant surfaces. Further, adults regularly feed on fruit that are ripe or nearly so, negating the option for treatment near harvest with most conventional insecticides. Thus, control strategies include using smoke to mask odor of ripening fruit; smoke is deployed just before dusk to several hours after nightfall. A labor-intensive but effective method requires bagging or screening fruit in the field; this method is only economically feasible when fruit are of high value and/or easily accessible. Attract-and-kill bait stations are also being developed, and several parasitoids have been identified as potential biological control agents.

Leafrollers

Leafrollers are moth species belonging to the family Tortricidae; they can be found in all deciduous tree fruit-growing regions and are considered to be serious pests of stone and pome fruit. Adults are small, 8 to 30 millimeters in length, with broad, brown forewings and wingspans of approximately 30 millimeters, depending on species. Eggs are small, cryptically colored, and laid in masses on the upper surfaces of leaves. Larvae hatch and feed within shelters of folded or rolled leaves attached by silken threads. Larval feeding often includes fruit surfaces, leaving behind damage in the form of tiny holes or tunneling. There are multiple generations each year, and overwintering stages are either larval or pupal depending on species. Monitoring with pheromone traps to capture adult males and scouting for larvae based on degree-day accumulations are useful tools for timing insecticide sprays. Organophosphate- and carbamate-based insecticides traditionally have been used for control, although insect growth regulators and other newer insecticide chemistries are now available and being used.

Oriental Fruit Moth

Oriental fruit moth, *Grapholita molesta* (Busck), is a pest of stone fruit, and more recently apples and pears, and is found throughout the world. This small grayish moth species is approximately 5 millimeters in length with a wingspan of 10 to15 millimeters and belongs to the family Tortricidae. Flattened, oval-shaped, lightly colored eggs are laid singly on undersides of leaves or on twigs and in later generations, directly on fruit. Larvae are often confused with those of the codling moth; they can be differentiated by the presence of an anal comb on the last abdominal segment of oriental fruit moth larvae (Figure I1.1). Full-grown larvae overwinter in silken cocoons on trees within bark crevices, beneath groundcover, in weeds, or in orchard trash. Larvae pupate the following year in spring, with adults being found between pink bud stage and bloom, depending on cultivar and location. There are multiple generations per year, with the number completed depending primarily on temperatures experienced in a geographic region. First-generation larvae bore into new growth stems, resulting in damage to terminals, referred to as "flagging," while in later generations, larvae bore directly into fruit to feed internally, leaving signs of injury that include exuded gum and frass. However, if larvae enter fruit from inside the stem, there may be no external evidence of entry or injury. Monitoring methods include at least two pheromone traps per orchard to capture adult males and the use of a degree-day egg hatch model based on a biotic point (biofix) of first sustained adult catch in pheromone traps. Organophosphate-based insecticide programs have been the basis of oriental fruit moth control, although mating disruption has also proven promising for this species in orchards of at least 2 hectares in size.

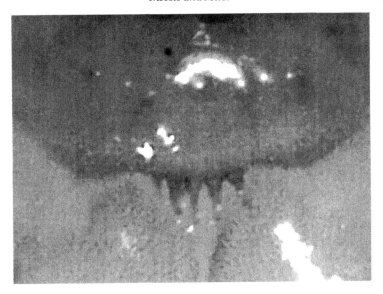

FIGURE I1.1. Anal comb (200x) on the last abdominal segment of an oriental fruit moth larva, differentiating it from a codling moth larva (*Source:* Courtesy of Henry W. Hogmire Jr., West Virginia University, Kearneysville, WV.)

Plant Bugs and Stink Bugs

Species of plant bugs (Family Miridae), especially *Lygus* species, and stink bugs (Family Pentatomidae) are pests of stone and pome fruit throughout the world. Although adults and nymphs feed on many herbaceous plants (especially legumes), they also will feed regularly on deciduous tree fruit and shoots. In stone fruit, feeding injury prior to shuck split results in flower and fruit drop, while feeding on fruit before pit hardening primarily causes catfacing injury. Later-season feeding results in additional surface blemishes, water-soaked areas, and gummosis. In apples, early season, prebloom feeding results in bud abscission, while feeding after fruit set results in slight dimpling to deeply sunken, distorted areas. White sticky, rectangular traps hung from trees have been used to monitor tarnished plant bug, *Lygus lineolaris* (Palisot de Beauvois), in apples, but not as successfully in stone fruit. Sweep sampling using a net in groundcover and limb jarring over a beating tray are good methods for sampling bugs in an orchard. Weed control, mowed grass, and clean cultivated aisles all help reduce both plant bug and stink bug populations. Prebloom insecticide sprays are often used to help control bugs early in the season.

INDIRECT PESTS

Borers

The most common species of bark and tree borers of pome and stone fruit include Coleopteran species belonging to the families Buprestidae, Cerambycidae, and Scolytidae and Lepidopteran species belonging to the family Sesiidae. Larvae damage trees by feeding on or beneath bark within the cambial layer of wood; feeding can take place in or on roots, trunks, or limbs, depending on species. Often, eggs are laid at sites of injury or in cracks on bark of trees. Signs of infestation include frass, sawdust, and gum. Larval feeding often facilitates entry of secondary insect and disease problems. Infested trees become less productive with steady declines in vigor and yield. Depending on the species and level of infestation, trees can be lost. Monitoring for Sesiidae species often involves the use of pheromone traps to capture female-seeking adult male moths. However, signs of frass, sawdust, and gum are also good indicators of infestation. Mating disruption is available as a control strategy for several Sesiidae species, but for most, insecticide sprays directed at the trunk and scaffold limbs are used for control. However, control may be difficult if larvae feed on roots or deep within wood.

European Red Mite

European red mite, *Panonychus ulmi* (Koch) (Family Tetranychidae), is a worldwide pest of tree fruit, with apple considered to be its most important host. Adult females are globular shaped with four rows of white hairs on their backs and change from brownish green to brownish red one to two days after molting. Adult males are smaller with a tapered abdomen and reddish yellow coloration. Eggs are orange (summer) or bright red (winter) with a long stalk protruding from the top. The six-legged larvae are pale orange but darken to pale green after feeding. Later nymphal stages have four pairs of legs and are green to reddish brown. The egg is the overwintering stage; eggs are deposited on twigs, and larvae hatch the following year to feed on foliage early in the season in apple, with adults appearing by petal fall. Up to ten generations can occur in a normal growing season. Foliar feeding results in bronzing of leaves, as chlorophyll and cell contents are destroyed. Moderate to severe infestations reduce yield by decreasing fruit size and promoting premature drop. Damaging populations also may lead to fewer and less vigorous fruit buds the following year. Monitoring involves scouting for eggs on twigs and spurs in the dormant season and foliar examinations throughout the growing season to determine the level of infestation with this species as well as twospotted spider mite. One or more prebloom oil applications along with miticide treatments have been used for control throughout the growing season. Natural enemies such as predaceous mite species and ladybird beetles can reduce populations if chemicals that are harmful to these beneficials are avoided.

Leafminers

Leafminer species belonging to the genus *Phyllonorycter* (Family Gracillariidae) can be found in fruit-growing regions of Europe, North America, and Asia and attack apples and pears, although cherries, quinces, plums, and crabapples are occasionally damaged. *Phyllonorycter* species are very small moths, 3 to 5 millimeters in length with 7 to 9 millimeter wingspans; forewings are heavily fringed and bronze colored with white and brown streaks. Eggs are clear to light green, flattened oval in shape, and attached to the undersides of leaves. Larvae are yellowish in color and feed within leaves between epidermal layers; early instars generally feed on sap or protoplasm of cell contents while later instars consume tissue. Pupae are brown and found within the leaf mines themselves. Pupae are the overwintering stage, and there are multiple generations per year. Pheromone and visual traps can be used to monitor flights of female-seeking adult males. However, visual inspection of leaves for mines is recommended for most species as well. Each mine can reduce effective leaf area by 4 to 5 percent. Heavy infestations can lead to complete defoliation and reduced crops and can adversely affect future fruit production. Biological control has proven effective, as 30 percent parasitism rates of first generation tissue-feeding larvae are considered to be great enough to provide effective control for the rest of the season. Insecticide application to control latter generations can reduce natural enemy populations and therefore should be avoided due to the potential for greater control problems the following year.

Pear Psylla

Pear psylla, *Cacopyslla pyricola* (Foerster) (Family Psyllidae), is the most important pest of pears worldwide. Adults are 2 to 3 millimeters in length; summer-form adults range in color from greenish orange to reddish brown while overwintering adults are darker reddish brown to black. Eggs are 0.5 millimeter in length, oval-shaped, and change from white to yellow before hatching. Young nymphs are pale yellow with red eyes, while older nymphs develop darker coloration and distinct black wing pads. Adults overwinter in and around pear orchards. They mate in spring, and females lay eggs on or near developing buds. After eggs hatch, nymphs move to axils of leaf petioles and to stems, where they feed on sap with sucking mouthparts and surround themselves with an ever-increasing drop of honeydew, eventually leading to growth of sooty mold. Multiple generations can occur each year. Heavy infestations can cause premature leaf drop, weakened fruit buds, and reduction in shoot growth. Monitoring begins while buds are still dormant by tapping limbs to look for adults. During the summer, terminal leaves are examined. Dormant oil sprays along with synthetic insecticides, especially pyrethroids and miticides, have been traditionally used for control. Newer chemistries, including insect growth regulators, are becoming available. However, cultural control strategies, such as summer pruning and limited nitrogen applications to reduce excessive tree vigor, can aid in population reduction. Many predatory insects will feed on pear psylla, and avoiding chemical treatments that harm these beneficial insects can aid in population reduction as well.

Rosy Apple Aphid

Rosy apple aphid, *Dysaphis plantaginea* (Passerini) (Family Aphidae), is found in most apple-growing regions worldwide. Oval-shaped eggs are laid on twigs and branches in the fall; coloration of eggs changes from bright yellow to jet-black within two weeks of oviposition. Eggs hatch into nymphs in the spring from the silver tip to one-half-inch green stage of bud development. Nymphs feed on leaves and fruit buds until leaves begin to unfold and change from a dark green to a rosy brown hue as they develop into adults. The mature adult, called a stem mother, reproduces asexually, and a second generation then matures on apple. Winged adults from third and sometimes fourth generations disperse to summer hosts such as narrow and broadleaf plantain. Multiple generations will reproduce on these hosts. Winged females return to apple trees in the fall to produce live female young who will mate with males and subsequently deposit four to six overwintering eggs. Nymphs inject a toxin as they suck sap that results in leaf curling, which may become severe and result in abscission. Furthermore, heavy infestations on apple lead to deformed and stunted fruit due to translocation of salivary toxins. Honeydew produced by colonies can lead to growth of sooty mold. Monitoring to determine the need for insecticide application requires visual inspection of clusters on susceptible apple cultivars to look for presence of aphids; monitoring should occur at pink and petal fall in apple. Predators such as ladybird beetles, syrphid flies, lacewings, and predatory midges, as well as parasitic wasps, are capable of providing effective biological control if chemicals toxic to these beneficial insects are avoided.

Twospotted Spider Mite

Twospotted spider mite (Family Tetranychidae) attacks tree fruit, ornamentals, and vegetable crops worldwide. Adults are pale and oval-shaped with two black spots located behind the eye spots; males generally are smaller than females. Eggs are clear and spherical and hatch into six-legged larvae that change from nearly transparent to dark green coloration after feeding. Later nymphal stages also are green but with four pairs of legs and more pronounced spots. Adults overwinter in orchards under bark or groundcover and become active in the spring as temperatures reach 12°C, when they disperse upward into orchard canopies to lay eggs. Females reproduce offspring of both sexes if mated, while unfertilized females produce males only. Shortly before the adult female emerges, she releases a pheromone to attract males and ensure female progeny. Up to nine generations occur throughout the season. Foliar damage caused by feeding is characterized by bronzing of leaves, as chlorophyll and cell contents are destroyed. Moderate to severe infestations reduce yield through decreased fruit size and lead to fewer and less vigorous fruit buds the following year. Regular foliar inspections can aid in determining if chemical treatment is necessary, based on percentage of leaves infested with twospotted spider mite and/or European red mite. Miticide treatments traditionally have been used to control this species and European red mite throughout the growing season. Natural enemies such as predaceous mite species and ladybird beetles can reduce populations if chemicals that are harmful to these beneficials are avoided.

BENEFICIAL INSECTS

Beneficial insects and mites, also known as natural enemies, reduce pest populations in orchard ecosystems via parasitism or predation, termed biological control. Parasitoids are smaller than prey and slowly kill them by developing as external or internal parasitic larvae. Predators are free-living beneficials that are as large or larger than their prey and kill and consume more than one prey item in their lifetimes. Use of beneficials can be classified into one of the following categories: conservation, augmentation, inundation, or introduction. Conservation involves creating favorable habitat for beneficials by reducing pesticide applications that harm beneficials and adding alternate food sources. Augmentation involves releases of mass-reared beneficials to bolster existing populations. Inundation also utilizes releases of mass-reared beneficials but with the goal to saturate the system and control pest populations within one generation. Introduction, or classical biological control, involves the release of an exotic beneficial to control a pest; this method is generally most effective when the pest is also an exotic member of an ecosystem and therefore has no effective natural enemy present. The most common approaches used in orchard ecosystems are conservation and augmentation.

Parasites

Major parasitoid groups found in conventional orchards include species of parasitic wasps belonging to the Braconidae, Ichneumonidae, and Eulophidae families as well as parasitic flies belonging to the Tachinidae family. Leafminer pest species can be controlled to acceptable levels in conventional orchards by parasitoids, and populations of other pests can be substantially reduced by their presence.

Predators

The most important predators found in conventional orchard ecosystems are generally associated with control of mite and aphid species. These include predaceous mites (Family Phytoseiidae) and ladybird beetles (Family Coccinellidae) (Figure I1.2) for control of mites. For control of aphids, the most common include fly species belonging to Syrphidae, Asilidae, and Cecidomyiidae families, green lacewings (Family Chrysopidae), as well as ladybird beetles.

Although insect and mite pests described here are of present-day importance on a global scale, new pests could emerge as management practices change, new insecticide and miticide chemistries are introduced, and new tree cultivars are planted. Furthermore, the potential impact of global climate change as well as introduction of exotic insects and mites to new regions also could lead to new pest problems.

Related Topics: DISEASES; PLANT-PEST RELATIONSHIPS AND THE ORCHARD ECOSYSTEM; SUSTAINABLE ORCHARDING

FIGURE I1.2. Ladybird beetle (Family Coccinellidae), an extremely important predator of aphids in orchard ecosystems (*Source:* Courtesy of Mark W. Brown, U.S. Department of Agriculture, Kearneysville, WV.)

SELECTED BIBLIOGRAPHY

Hogmire, H. W. Jr., ed. (1995). Mid-Atlantic orchard monitoring guide, Publication NRAES-75. Ithaca, NY: Northeast Regional Agric. Engin. Serv.

Howitt, A. H. (1993). *Common tree fruit pests,* NCR 63. East Lansing, MI: Michigan State Univ. Exten. Serv.

McPherson, J. E. and R. M. McPherson (2000). *Stinkbugs of economic importance in America north of Mexico.* New York: CRC Press.

Penman, D. R. (1976). Deciduous tree fruit pests. In Ferro, D. N. ed., *New Zealand insect pests* (pp. 28-43). Canturbury, New Zealand: Lincoln College of Agric.

Sands, D. P. A., W. J. M. M. Liebregts, and R. J. Broe (1993). Biological control of the fruit piercing moth, *Othreis fullonia* (Clerck) (Lepidoptera: Noctuidae) in the Pacific. *Micronesia* 4:25-31.

Travis, J. W., coordinator (2000). *Pennsylvania tree fruit production guide.* State College, PA: Pennsylvania State Univ. College of Agric.

Van Der Geest, L. P. S. and H. H. Evenhuis, eds. (1991). *Tortricid pests: Their biology, natural enemies and control.* Amsterdam, the Netherlands: Elsevier.

Van Driesche, R. G. and T. S. Bellows (1996). *Biological control.* New York: Chapman and Hall.

– 2 –

Irrigation

D. Michael Glenn

Irrigation is required for producing deciduous tree fruit crops when there is inadequate water available in the soil from precipitation to meet the atmospheric demand for water through the tree. Climate is generally the primary indicator for irrigation need. Arid regions with less than 250 millimeters of rainfall require irrigation even for tree survival; semiarid regions receiving less than 500 millimeters can produce a fruit crop, olive for example, but yields are low and the risk of crop failure is high. Subhumid and humid regions receiving more than 500 millimeters of rainfall can produce fruit crops, depending on the available water storage capacity of the soil. Irrigation is often supplied in subhumid and humid regions when shallow or sandy soils limit the available water storage in the soil or drought frequently occurs for periods greater than two weeks. Irrigation should be supplied to newly planted and young trees when their root systems are poorly developed.

IRRIGATION SCHEDULING

Irrigation scheduling requires knowledge of two crop characteristics: (1) how much water a tree needs and (2) when it should be applied. The amount of water a tree crop uses is called evapotranspiration (ET) and is based on both the atmospheric demand for water and the ability of the soil to supply it. Potential ET refers to the maximum ET rate from a large area covered completely and uniformly by actively growing vegetation with adequate moisture at all times. Potential ET is generally determined using computer models that utilize weather data consisting of solar radiation, temperature, wind speed, and relative humidity. Potential ET can also be estimated with a class A evaporation pan or a potometer. Based on tree age, height, species, and crop load, crop coefficients are used to adjust the potential ET to the actual ET that must be applied in irrigation. For example, a young apple orchard with an incomplete canopy within the row might have a crop coefficient of 0.70, whereas a fully mature orchard may have a crop coefficient of 1.25. Crop coefficients are time and locale specific, so local extension and other agriculture resources should be consulted for explicit information. Another approach to determining how much water an orchard needs is to measure water use from the soil and replace the same amount through irrigation. Soil moisture sensors and sensor access tubes can be installed in the root zone of the orchard and monitored periodically. These sensors will determine how much water has been removed from the root zone through actual ET, and the same amount is replaced through irrigation.

Irrigation timing depends on both the plant requirements for water and the capacity of the irrigation system to supply water. In general, deciduous fruit trees can tolerate a reduction of 50 percent of the available water in the root zone before economic stress levels occur. Available soil water is the amount of water retained in the soil between field capacity and the permanent wilting point. Field capacity can be estimated as the amount of water in the soil one to three days after a full irrigation or a prolonged period of rain. The permanent wilting point is the amount of water that remains in the soil when plants are no longer able to transpire. The permanent wilting point is difficult to measure in the field; however, for a wide range of soil types, it is approximately 50 to 75 percent of the field capacity value. On shallow or sandy soils, 50 percent depletion of available water can occur in less than five days, while on deep silt loam soils, the water-holding capacity of the soil can provide adequate water for up to 14 days in many climates. Young trees planted on any soil type have a limited root zone and will require frequent irrigation in the absence of frequent and effective precipitation.

Irrigation systems are generally designed to provide water, in rotation, to numerous sections of an orchard. Less water reserves are needed for a single irrigation event, and a smaller pump using less energy can be used. The irrigation pump size and the number of sections in the orchard are initially designed to ensure that when the orchard is mature, sufficient water can be supplied to all sections under the maximum ET demand for the region. Irrigation design is a complex discipline.

IRRIGATION SYSTEMS

There are three major irrigation systems in tree fruit production. These include surface irrigation, overhead or sprinkler irrigation, and microirrigation.

Surface Irrigation

Surface irrigation utilizes evenly spaced channels, or furrows, to direct free-flowing surface water into the basin or field. The land must be sloped, and the water enters the field on the high end and flows to the low end where the excess is collected and returned to a distribution ditch. Surface irrigation has the following advantages: (1) high application rates, (2) low capital investment, and (3) effectiveness on soils with surface crusting. Limitations and disadvantages include (1) the potential for excessive soil erosion, (2) concentration of salts on the furrow ridges through evaporation of water from the soil, (3) ineffectiveness on sandy or coarse-textured soils, (4) subsurface water loss, (5) high water loss from evaporation, and (6) a requirement for land leveling.

Sprinkler Irrigation

Sprinkler irrigation systems are permanently installed or have moveable irrigation distribution lines, laterals, and risers with sprinkler heads. In the moveable systems, aluminum pipes or flexible plastic tubing are moved with the sprinklers from one location to another. In permanent systems, the primary distribution pipes are buried, and only the risers and sprinkler heads are aboveground. Sprinkler irrigation is adapted to a wide range of soils and topographies. Sprinkler spacing varies from 5 by 5 meters to 73 by 73 meters, with the output of each sprinkler head and pressure increasing as the spacing is increased. In orchard systems, closely spaced sprinklers are most common. Sprinkler heads use pressure energy to break the flow of water into smaller droplets that are distributed over the land and crop. They are located either below the tree branches or above the tree canopy to cool the tree through evaporative cooling. A modification of sprinkler irrigation is microsprinkler technology in which microsprinkler heads are permanently located between each tree and below the canopy in the tree row. Microsprinkler heads deliver a more frequent and lower volume of water than conventional sprinkler heads. A well-designed sprinkler irrigation system has a high uniformity of water distribution and can deliver either frequent and low-volume applications that meet daily water needs or infrequent and high-volume applications that meet seven- to 14-day water requirements.

Sprinkler irrigation has many advantages. It can be used on permeable soils with rolling topography not conducive to surface irrigation and is adaptable to all irrigation frequencies. In arid and semiarid regions, tree rows and grass driveways can be irrigated together. Application rates can be accurately controlled to minimize subsurface water loss, and water with moderate levels of sediment can be used. Two advantages with great economic impact are that it can be designed for frost protection under radiation frost conditions or for blossom delay in areas with a high probability of spring frost. When plants are coated with water, the heat of fusion of the water freezing maintains temperatures near 0°C, rather than allowing the plant to reach temperatures many degrees below freezing. Deciduous fruit flowers can survive temperatures at freezing, but the ice coating on the trees must be continually sprayed with water until it melts. Application rates for frost prevention range from 2 to 7 millimeters per hour, depending on temperature and wind conditions. Floral emergence can be delayed as much as 14 days with frequent wetting applications when air temperature is generally above 5°C. Maximum cooling is achieved with an automatically programmed irrigation system that schedules irrigation wetting based on air temperature. Another benefit of sprinkler irrigation is that it can be used for crop cooling and heat stress reduction to improve yield, color development, and internal fruit quality. When plants are coated with water, the latent heat of vaporization of water evaporation cools the wetted surface up to 14°C. Scheduling of crop cooling is based on air temperature and relative humidity and requires an automatically programmed irrigation system. A final advantage of sprinkler irrigation is that fertilizers, pesticides, and plant growth regulators can be applied using the water distribution system. Coverage on the underside of leaves, however, is generally poor so the most effective materials are those which are absorbed into the plant. Disadvantages of sprinkler irrigation are (1) moderate capital investment in wells, pumps, and water distribution lines; (2) tree and fruit damage from water with high salt content applied to the canopy; (3) increased likelihood of disease development, if applied to the canopy, requiring more fungicide usage; and (4) high water loss from evaporation of the sprayed water.

Microirrigation

Microirrigation, formerly known as drip or trickle irrigation, systems are permanently installed water distribution designs that deliver frequent, low-volume water applications to the soil along the distribution lines, generally through pressure compensated emitters. The water moves into the soil and spreads primarily through unsaturated water flow. The volume of soil wetted by each emitter and the number of emitters per plant are determined by both the flow and frequency of irrigation as well as the soil hydraulic properties. The number of emitter points providing water to an individual tree varies from 0.5 to 4.0 and can be increased as a tree grows. Traditional microirrigation systems are placed on the soil surface or hung from a trellis wire above the soil. A modification of microirrigation systems is the subsurface irrigation system, which is permanently buried below tillage depth and generally 0.3 to 1.0 meters deep within the tree row. Microirrigation has the following advantages: (1) improved penetration on problem soils, due to application of water at low rates; (2) reduced salt accumulation and more dilute and less phytotoxic salt concentrations in the soil water, due to frequent water applications; (3) efficacy with saline water sources; (4) reduced soil surface evaporation, runoff, percolation losses, and weed growth (in the non-irrigated areas), since less than 100 percent of the root zone is irrigated; (5) uninterrupted cultural operations, e.g., weed and pest control applications, since only a portion of the orchard is irrigated; (6) adaptability for applying nutrients; and (7) savings in water and pumping costs due to improved water use efficiency. Disadvantages of microirrigation systems include (1) surface and subsurface damage of distribution lines by animals and farming operations; (2) the inability to supply water to grass drive middles; (3) high capital costs for pumps, filters, distribution lines, and emitters; and (4) emitter plugging. Water filtration and chemical treatment of water quality are critical in all microirrigation systems and absolute requirements in subsurface systems. Plugging may occur from sediment in the water or chemical reactions of water in the distribution lines and the emitter openings. Biological growth of microorganisms in the distribution lines can also cause emitter plugging. A major problem of subsurface irrigation is the intrusion of plant roots into emitter openings.

Irrigation is a necessary part of deciduous tree fruit production throughout the world and is a key component in providing a stable and high-quality product for the marketplace. The competing demands for water from urban areas, industry, and recreation sources, in addition to the degradation of water sources from salinity, erosion, and overuse drive agriculture to find more efficient ways of providing water to tree crops and to make the most efficient use of natural resources.

Related Topics: ORCHARD PLANNING AND SITE PREPARATION; SPRING FROST CONTROL; TREE CANOPY TEMPERATURE MANAGEMENT; WATER RELATIONS

SELECTED BIBLIOGRAPHY

Lamm, F. R., ed. (1995). *Microirrigation for a changing world: Conserving resources/preserving the environment, Proceedings of the fifth international microirrigation congress.* Orlando, FL: Amer. Soc. of Agric. Engin.

Microirrigation forum: A comprehensive source of irrigation information (nd). Retrieved September 1, 2001, from <http://www.microirrigationforum.com>.

Williams, K. M. and T. W. Ley, eds. (1994). *Tree fruit irrigation: A comprehensive manual of deciduous tree fruit irrigation needs.* Yakima, WA: Good Fruit Grower.

L

1. LIGHT INTERCEPTION AND PHOTOSYNTHESIS

– 1 –

Light Interception and Photosynthesis

David C. Ferree

A number of summary papers and review articles (Flore and Layne, 1996; Jackson, 1980; Lakso, 1994) cover photosynthesis, source-sink interactions, and light relationships of temperate zone tree fruit. Research work on apple is the most comprehensive; therefore, most of the following examples and discussion focus on this work, supplemented with results from other crops.

PHOTOSYNTHESIS

Temperate zone fruit trees have the C_3 photosynthetic pathway and utilize light as the energy source to drive the process. The photosynthetic light response curve is hyperbolic. The rate of photosynthesis of a leaf increases rapidly until it is saturated around 30 percent full sunlight (700 to 1,000 micromoles quanta per square meter per second photosynthetically active radiation). The light compensation point occurs around 20 to 50 micromoles quanta per square meter per second. Light levels above the saturation level have little direct effect on photosynthesis, as other factors become limiting. Several studies show that the areas of the tree canopy that receive 30 percent full sunlight also are the areas that readily initiate flowers and have the largest fruit and highest fruit quality. Because of shading differences in leaf angle, canopy density, and canopy form and shape, the whole leaf canopy likely never becomes photosynthetically saturated.

Fruit tree leaves reach their maximum rate of photosynthesis early in the growing season and become net exporters of photosynthate when they are 10 to 40 percent fully expanded. Early development of apple leaves combined with slow leaf aging and decline of photosynthesis translate into a long duration of effective photosynthetic leaf area. This characteristic may form the basis for high yield potential. Results from whole tree studies demonstrate the same general pattern of a high rate of photosynthesis during most of the growing season, declining in the fall around harvest. Apple is unique in having primary spur leaves that form the early canopy and are the sources of photosynthate during cell division. Later in the season, after fruit set, bourse shoot and terminal shoot leaves become the primary sources supplying the fruit with carbohydrates. Spur leaf area is closely related to fruit set, fruit size, fruit soluble solids, long-term yield, and fruit calcium level at harvest (Ferree and Palmer, 1982). Recent work by Wunsche and Lakso (2000) shows that fruit yield is linearly and highly correlated ($r^2 = 0.78$) with spur leaf light interception in a range of different orchard systems.

Early in the season, leaves on spurs with fruit have higher levels of photosynthesis than spurs without fruit, while late in the season this trend is reversed. Leaf efficiency, in a number of studies, is closely related to leaf photosynthetic rate and previous and present light environment. Spur leaves supply the fruit with photosynthate early in the season during cell division, while shoot leaves supply the fruit later in the season during cell expansion. Evaluations in various orchard management systems indicate that spur leaf efficiency follows light level in the canopy. Shoot leaves have higher rates of photosynthesis than spur leaves and also have larger leaf areas.

Apple leaves adjust very quickly to changes in light pattern, and studies on intermittent lighting indicate that apple leaves may be about 85 percent as efficient under alternating light levels as when steady-state light is provided (Lakso, 1994). A single leaf can register multiple levels of photosynthesis. The portion of a leaf in a sunspot will be much higher than the remainder of the leaf that may be in the shade. These characteristics make an apple tree very responsive to the rapid changes in light distribution during the day and also adaptive to longer-term changes that may occur as the crop weight causes limb repositioning.

The skin of young apple fruit photosynthesizes but its contribution to the overall photosynthate of the tree is very small compared to the leaf contribution. The role of fruit photosynthesis may be greater in peach and cherry. Research by Flore and Layne (1996) demonstrates that fruit gross photosynthesis contributes 19.4, 29.7, and 1.5 percent of the carbohydrate used by the fruit during Stages I, II, and III, respectively.

FACTORS AFFECTING PHOTOSYNTHESIS

Environmental Factors

Although it is obvious that light affects photosynthesis, soil moisture and temperature also have a role. Water stress not only reduces photosynthesis per unit leaf area but also reduces leaf size, so overall capacity is reduced. In extreme situations, premature leaf abscission occurs. Stomates can close in response to low humidity. Wind can alter leaf water loss by affecting the boundary layer and by causing an increase in transpiration. Excess soil moisture affects trees by reducing oxygen to the roots and possibly by increasing the accumulation of soil carbon dioxide. Apple transpiration, photosynthesis, leaf growth, and root growth can be reduced by several days of flooding. The sensitivity to flooding appears to be related to periods of most active growth.

Lakso (1994) reports that photosynthesis of apple leaves does not have a strong response to temperatures over a fairly wide range—from 15 to 35°C, with the optimum generally near 30°C. Temperatures of 37°C and above are deleterious to photosynthesis. Frosts in the fall decrease photosynthesis and, if severe, will kill the leaves. Growing areas with long periods of frost-free conditions following apple harvest tend to have higher fruit yields and size because of increased storage reserves and bud development during this period.

Cultural Factors

A number of cultural factors can influence photosynthesis of tree fruit. Nutrient deficiencies of several elements, particularly nitrogen, cause decreases in photosynthesis. Increasing concentrations of ozone are reported to cause a linear decrease in photosynthesis of almond, apple, apricot, pear, plum, and prune, while cherry, peach, and nectarine are unaffected (Retzlaff, Williams, and DeJong, 1991). Certain pesticides, particularly spray oils and the organically approved materials copper and sulfur, cause a reduction in photosynthesis of apple leaves (Ferree, 1979). Research in Ohio shows that mites, apple scab, and powdery mildew also reduce apple leaf photosynthesis. In a simulated leaf injury study, losses of leaf area up to 7.5 percent have no significant effect on apple leaf photosynthesis, but significant reductions occur when losses of leaf area exceed 10 percent. If losses exceed 15 percent, not only is the photosynthetic capacity of the leaf reduced, but the performance of the remaining leaf tissue is also reduced. This reduction is associated with injury to veins and increased circumference of the simulated insect feeding. Thus, injury that results in many small holes is more serious than equivalent area lost in a few large holes. Similar results are reported on cherry.

Although differences among apple cultivars grown under similar conditions are small, spur-type 'Delicious' appears to have higher photosynthetic rates than standard habit 'Delicious'. The effect of rootstock on photosynthesis is nonexistent in some studies, but in others, photosynthetic rates of trees on more vigorous rootstocks are greater than on dwarfing rootstocks. Summer pruning delays the natural decline in photosynthesis of the remaining leaves. Root pruning, which acts mostly through interruption of water relations, causes a temporary reduction in photosynthesis. Orchard management and training systems have little effect on apple leaf photosynthesis as long as similar leaf types and exposures are measured.

LIGHT INTERCEPTION

Two kinds of light are important—direct and diffused, or indirect. Direct light is uninterrupted as it falls directly on the leaf, while diffused light is reflected from clouds, particles in the air, or leaves. Shade of a single leaf reduces the light level well below the saturation level for photosynthesis. As light travels through a tree canopy, much of the visible range is absorbed, and thus the spectral balance in the lower portion of the canopy is relatively higher in infrared than the canopy periphery. Recent evidence suggests that some of the differences in growth response and leaf shape in the interior of the canopy may be due to these changes in spectrum. Lakso (1994) reports that light levels on the interior of the canopy will be higher on hazy or partly cloudy days than on clear days or very overcast days. Some fruit-growing areas, such as Italy, New Zealand, and the West Coast of the United States, have 30 to 40 percent more light than the eastern

United States or western Europe. Light also declines at more northerly latitudes and results in lower yields and smaller fruit size.

The sun provides a point source of light, and a classic work conducted by D. R. Henicke in the 1960s shows that a shell over the top and partially down the sides of a tree receives light in excess of saturation. An inner shell receives an adequate level and the shell in the bottom and center of the canopy receives inadequate light to saturate photosynthesis. In the old-fashioned, large seedling tree, the portion of the canopy receiving inadequate light may be as much as 30 to 40 percent of the total canopy. One of the greatest advantages of dwarf trees is that this interior shaded area is greatly reduced on each tree, and with many more trees per hectare, light interception and orchard efficiency are dramatically increased. Transmittance of light through an apple leaf is about 7 percent of photosynthetically active radiation, based on research by Jackson (1980). Reflectance is greater than transmittance over all visible wavelengths and is greatest in May.

RESPONSES OF FRUIT TREES TO LIGHT

Yield

Lakso (1994), in a summary of many studies conducted over a 30-year period, demonstrates clearly that as total light interception increases, apple yield per unit of land increases. The implication is that areas of inherently high light levels will have higher yields than areas of lower light levels. Interception levels above 50 percent of available light result in more variability in the data, indicating other factors become more important. Another major factor affecting yield is how light is distributed through the canopy and how much of the canopy is above the critical level of light needed to saturate photosynthesis, cause flower initiation, and foster cell division for early fruit growth. Shaded areas of the canopy (less than 30 percent full sun) result in smaller fruit size and increased preharvest drop. With peach, shade has its greatest effect on fruit weight and quality during Stage III of development.

Vegetative Response

Leaves that develop in the sun have higher nitrogen levels, higher specific leaf area, more palisade layers, smaller leaf area, more cupping or curling, and less chlorophyll. Generally, shoot growth is greater in areas with high light and a long season. Trees under these conditions have greater precocity. Marini, Sowers, and Marini (1991) report that minimum light threshold for peach shoot growth is lower than that required for flower initiation, which is the opposite of the requirements for apple.

Flowering

Apples are considered day neutral, and it does not appear that flower initiation, fruit growth rate, fruit size, or fruit color are phytochrome mediated. However, apples contain phytochrome, and fruit set and preharvest fruit drop can be influenced by changes in the red:far-red (R:FR) ratio mediated through phytochrome. Studies show that flower initiation does not occur in canopy areas receiving less than 30 percent full sun for apple and 20 percent full sun for peach and cherry.

Fruit Color

Another direct response to light is fruit color. Anthocyanin formation in apple requires light, and a single leaf lying on the shoulder of an apple prevents red color formation. The poorest-colored and smallest fruit come from the most shaded portions of the canopy. On most cultivars, 30 to 50 percent full sun is needed to ensure adequate red color development and good fruit size. There is evidence that some highly colored 'Delicious' strains can be fully colored with as little as 9 percent full sun. However, soluble solids and starch levels in fruit receiving low light levels are far inferior to those in fruit from well-illuminated portions of the canopy. Fruit size and shape are always best in sections of the canopy that receive 30 percent full sun or above.

ORCHARD PRACTICES TO IMPROVE LIGHT

Row Orientation

One of the simplest practices an orchardist can use to improve light is to orient rows north-south (Jackson, 1980). The north-south orientation provides a more even distribution of light over the canopy and, thus, more efficient production. The effect of row orientation is greatest on more upright training systems, such as the vertical axis, with much less effect on low, flat systems, such as the Lincoln canopy.

Pruning

Pruning is the orchard practice most consistently used to improve canopy light conditions. Studies indicate that pruning should be focused on improving light distribution and removing wood in unproductive positions on the limbs. Work at Ohio State University shows that mechanical or summer pruning can improve canopy light penetration and result in improved fruit color and soluble solids.

Reflectors

One of the artificial methods to increase light, particularly in the lower canopy, is to place reflective film on the alleyways between rows. Reflectors improve red color and in some instances, fruit size and soluble solids, particularly of fruit in the lower canopy. This may be a useful practice on hard-to-color, high-value cultivars but will be successful on only well-pruned trees with open canopies that allow light to penetrate to the reflector.

Orchard System

An orchard design combines a sequence of orchard practices into a coherent system to optimize light interception and distribution. One of the most efficient methods to decrease the portion of the canopy receiving inadequate light is to decrease tree size and plant more individual trees per hectare, thus greatly increasing well-exposed and productive canopy. Training systems such as the various trellis forms and the slender spindle capitalize on this principle. Tree shapes in which the bottoms are wider than the tops increase the percentage of

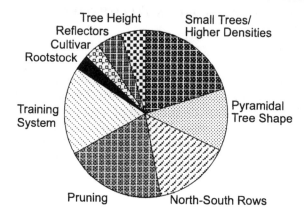

FIGURE L1.1. Orchard cultural practices to manage light

well-exposed canopy. Tree height becomes a concern if one row begins to shade the adjacent row. The most efficient method of reducing tree size is by size-controlling rootstocks. In addition to affecting tree size, some rootstocks impart an open, spreading character to the canopy that further improves light penetration. Some cultivars also influence canopy density.

There are many ways that light can be managed in fruit tree orchards. Relative importance of each is estimated in Figure L1.1. Light is the most important factor determining yield, fruit size, and quality, and its optimum management equates to improved orchard efficiency.

Related Topics: CARBOHYDRATE PARTITIONING AND PLANT GROWTH; DWARFING; FLOWER BUD FORMATION, POLLINATION, AND FRUIT SET; FRUIT COLOR DEVELOPMENT; FRUIT GROWTH PATTERNS; HIGH-DENSITY ORCHARDS; ORCHARD PLANNING AND SITE PREPARATION; TRAINING AND PRUNING PRINCIPLES; TRAINING SYSTEMS

SELECTED BIBLIOGRAPHY

Ferree, D. C. (1979). Influence of pesticides on photosynthesis of crop plants. In Marcelle, R., M. Clijsters, and M. VanPoucke (eds.), *Photosynthesis and plant development* (pp. 331-342). The Hague, the Netherlands: W. Junk.

Ferree, D. C. and J. W. Palmer (1982). Effect of spur defoliation and ringing during bloom on fruiting, fruit mineral level, and net photosynthesis of 'Golden Delicious' apple. *J. Amer. Soc. Hort. Sci.* 107:1182-1186.

Ferree, D. C. and I. Warrington (eds.) (2003). *Apples: Botany, production and uses.* Portland, OR: Oxford University Press.

Flore, J. A. and D. R. Layne (1996). Prunus. In Zamski, E. and A. A. Schaffer (eds.), *Photoassimilate distribution in plants and crops source-sink relationships* (pp. 825-850). New York: Marcel Dekker, Inc.

Jackson, J. E. (1980). Light interception and utilization by orchard systems. In Janick, J. (ed.), *Horticultural reviews,* Volume 2 (pp. 208-267). Westport, CT: AVI Publishing Co.

Lakso, A. N. (1994). Apple. In Schaffer, B. and P. C. Anderson (eds.), *Handbook of environmental physiology of fruit crops,* Volume 1 (pp. 3-42). Boca Raton, FL: CRC Press, Inc.

Marini, R. P., D. Sowers, and M. M. Marini (1991). Peach fruit quality is affected by shade during final swell of fruit growth. *J. Amer. Soc. Hort. Sci.* 116:383-389.

Retzlaff, W. A., L. E. Williams, and T. M. DeJong (1991). The effect of different atmospheric ozone partial pressures on photosynthesis and growth of nine fruit and nut tree species. *Tree Physiology* 8:93-103.

Wunsche, J. N. and A. N. Lakso (2000). The relationship between leaf area and light interception by spur and extension shoot leaves and apple orchard productivity. *HortScience* 35:1202-1206.

1. MARKETING

– 1 –

Marketing

Desmond O'Rourke

Marketing is a key activity for any commercial fruit enterprise. It involves both successfully transferring the product from the producer to a distant consumer and ensuring the flow of payment for the product back to the producer. While activities in the orchard or packing shed can influence successful marketing, most producers are dependent on a wide array of intermediaries to carry out other marketing functions. However, producers must bear the ultimate responsibility for choosing and monitoring those marketing agencies and for seeing that the consumer is satisfied and their own work is adequately rewarded. The competition for the consumer's favor is unrelenting from other fruit, other produce items, and other food and beverage offerings.

A number of frameworks can be used to analyze the marketing system for a particular product. In the case of temperate zone tree fruit, one common approach is to examine the key agents involved in the marketing system. A second is to analyze the key functions performed. Given the dynamic nature of the fruit marketing system, each approach brings a different light to bear on marketing challenges and business opportunities.

KEY AGENTS

The key agents involved in marketing temperate zone tree fruit are packers, processors, storage houses, shippers, marketers, promotional agencies, transportation companies, brokers, wholesalers, exporters, importers, retailers, and restaurants. Other entities such as banks, insurance companies, information providers, government inspection services, etc., help facilitate marketing but are not normally considered part of the marketing effort. The names of the agents are relatively self-explanatory but, in specific cases, may provide only limited information on what an agent actually does. Terms such as "shipper" or "marketer" are often used interchangeably. Many agents have integrated operations, for example, combining producing and packing or wholesaling and retailing. Brokers may act on behalf of buyers only or sellers only or exclusively for a few larger firms. Retailers may be direct importers for their own account. In practice, it is important to check what the agent actually does.

In most temperate zone tree fruit, regional or national promotional agencies play a pivotal role in the marketing of the product. Funds for promotion are collected from producers, packers, exporters, or government sources and entrusted to entities such as the Washington Apple Commission (apples) or the Chilean Fresh Fruit Association (multiple fruit). These entities coordinate activities such as sales representation, merchandising, promotion, or category management. They collaborate with the marketers, exporters, or importers who handle the actual product transactions. They provide a flow of information back to the producer on how market needs are changing and how the marketing system is adapting to those changes.

KEY FUNCTIONS

Most of the key functions involved in marketing fruit are also relatively self-explanatory. They include assembly, packing, processing, storage, transportation, exporting, importing, wholesaling, retailing, promotion, and category management. Assembly, packing, and processing normally occur in the producing district. There has been constant pressure in recent decades to increase the efficiency of these operations by increasing their size without compromising quality. Storage is crucial at every step of the marketing system for perishable products. There have been major advances in the speed of domestic and international transportation and in quality control of perishables while in transit, but workers and consumers still poorly understand how to keep fruit in optimal condition.

Exporting, importing, wholesaling, and retailing involve changes in the responsibilities of moving the product nearer to the consumer.

These entities evolved to fit past modes of transportation, price setting, and location of consumers. As society has changed, chain retailers have tended to become larger and have encroached more on the traditional business of the other traders. Since a large supermarket may carry 60,000 items, including 500 produce items, retailers are increasingly using supplier corporations, such as Dole or Chiquita, or industry associations, such as the Washington Apple Commission, as category managers. The category manager is given responsibility for stocking, display, and pricing of its specialty product in one or more retail divisions. Although retailers are willing to use local suppliers in retail outlets in their producing district, the competition for the retailer's business is increasingly global.

For most of the twentieth century, the marketing system for temperate zone tree fruit permitted the same product to pass through the hands of many independent actors of various sizes carrying out many overlapping functions. The system provided the producer and the consumer with many options. Product was always available on the open (spot) market. Transactions were numerous, and both price and volume reports were readily available. However, the system also led to excess costs, price volatility, and variable quality.

In the last two decades of the twentieth century, discount retail chains, such as Wal-Mart in the United States, Metro in Germany, and Marks and Spencer in Britain, attempted to reduce procurement costs by rationalizing the links in the supply chain for nonfood products. They used their purchasing power and global information systems to acquire large volumes of product, designed to their exacting specifications, and at low prices. This approach was so successful that it eventually was applied to prepared foods and, in the early twenty-first century, to perishables. Discounters grabbed a growing share of the global food market and forced traditional food chains to adapt similar methods in order to survive. Most large food retailers are now concentrating their purchases with fewer, larger suppliers. In many cases, relationships with suppliers are based on legal alliances and short- to medium-term contracts. Retailers are also increasingly requiring assurances on sustainable farm practices, food-handling methods, treatment of workers, etc.

The concentration of greater power at the retail level is having dramatic impacts on all other actors in the supply chain. The spot market has been considerably weakened. Processors, packers, and shippers have been consolidating to get large enough to compete for the giant retailer's business. They have been reexamining their relationship with producers in terms of how it affects their efficiency and cost. The consensus is that in order to remain a preferred supplier to the major retailers, their supplying growers will have to be able to produce greater volumes of better-quality products at lower costs. Fruit growers, too, must adapt to the market imperative. Entire producing districts could disappear in the shakeout.

While the mainstream marketing system for tree fruit is becoming more concentrated, there are still numerous niche opportunities for unconventional marketing. In most major cities, small, local retail chains survive alongside their giant competitors by utilizing their superior knowledge of local consumers' needs. On-farm or roadside direct marketers of fruit continue to thrive near major cities and highways by offering consumers a combination of quality fresh products with a rural experience. The demand for organic products was once confined to small specialty food stores but is now employed by health food chains, such as Whole Foods Markets, and in the produce section of conventional supermarkets. Upscale restaurants have also become major buyers of organic products. Community-driven marketing, where consumers contract with farmers for regular baskets of produce, are popular in some areas. A number of different models of electronic trading in produce are competing for participant suppliers and customers.

By its nature, marketing is dynamic. Competitors must constantly seek to stay ahead of their rivals by offering consumers improvements in price, quality, ambiance, or other psychic satisfactions. Throughout the supply chain, marketing agents constantly seek that innovation which will give them a temporary advantage. In such an environment, the most effective strategy for producers and suppliers of temperate zone tree fruit is to offer their own innovations in marketing. At the very least, they need to be aware of the market changes and take defensive measures to adapt to the new situations.

Related Topics: PACKING; PROCESSING; STORING AND HANDLING FRUIT

SELECTED BIBLIOGRAPHY

Alston, J.M., J.A. Chalfant, and N.E. Piggott (2000). The incidence of the costs and benefits of generic advertising. *Amer. J. Agric. Econ.* 82:665-671.

Colyer, D., ed. (2000). *Competition in agriculture: The United States in the world market.* Binghamton, NY: The Haworth Press.

Cotterill, R.W. (1996). The food distribution system of the future: Convergence towards the U.S. or U.K. Model. *Agribusiness, an International Journal* 12:123-135.

Kinsey, J. (2000). A faster, leaner supply chain: New uses of information technology. *Amer. J. Agric. Econ.* 82:1123-1129.

McCluskey, J.J. and A.D. O'Rourke (2000). Relationships between produce supply firms and retailers in the new food supply chain. *J. of Food Distribution Res.* 31:11-20.

O'Rourke, A.D. (1994). *The world apple market.* Binghamton, NY: The Haworth Press.

Sexton, R.J. (2000). Industrialization and consolidation in the U.S. food sector: Implications for competition and welfare. *Amer. J. Agric. Econ.* 82:1087-1104.

Hand packing apples on a commercial packing line. Hand packing machine-sorted fruit into shipping and retail display cartons is a strenuous and labor-intensive operation. (*Source:* Photo courtesy of Stemilt Growers, Wenatchee, WA.)

Robots on a commercial packing line removing 'Golden Delicious' apples from staging area and placing them on trays in boxes. New robotic equipment handles the delicate packing operation at speeds approaching 1.5 boxes per minute. (*Source:* Photo courtesy of Stemilt Growers, Wenatchee, WA.)

Mating disruption dispenser under test in peach orchards for oriental fruit moth control. Alternative production systems that view an orchard as a potentially sustainable agroecosystem are becoming more widely accepted as management strategies are developed that lead to less reliance on external inputs. (*Source:* Photo courtesy of Henry W. Hogmire Jr., West Virginia University, Kearneysville, WV.)

Comparison of apples treated (left) versus untreated (right) with a reflective material. Reflectants utilize the physical properties of mineral materials to modify the microclimate of a fruit tree to reduce water stress, sunburn, and other temperature-related injuries. (*Source:* Photo courtesy of D. M. Glenn, U.S. Department of Agriculture, Kearneysville, WV.)

Experimental planting comparing Lincoln canopy (foreground), slender spindle (middle), and horizontal suspended canopy (background) training systems. Training systems offer specific opportunities, and growers can choose from designs that increase productivity, improve fruit color and quality, augment integrated production efforts, boost work efficiency, and/or complement market strategies. (*Source:* Photo courtesy of Henry W. Hogmire Jr., West Virginia University, Kearneysville, WV.)

Spur-type 'Delicious' trees on M.9 rootstock trained to slender spindle system. Tree spacing is 1 meter by 3 meters. Orchards have undergone significant changes due to the widespread use of dwarfing rootstocks. (*Source:* Photo courtesy of Henry W. Hogmire Jr., West Virginia University, Kearneysville, WV.)

Peach trees with genetically improved tree forms. Combining novel tree traits that have potential applications for high-density fruit production with high fruit quality requires the collaboration of the fruit breeder with horticulturists specializing in tree training and production systems to effectively evaluate the complex traits leading to efficient fruit production for the grower and the qualities of the fruit that lead to consumer acceptance. Before large-scale testing of columnar (PI) and upright (UP) growth forms as compared with standard (ST) trees could begin, generations of breeding were required to combine these new columnar and upright tree forms (A) with high fruit quality (B). (*Source:* Photos courtesy of Ralph Scorza, U.S. Department of Agriculture, Kearneysville, WV, and Daniele Bassi, University of Milan, Italy.)

Disease-resistant pears and plums from USDA breeding programs. Combining disease resistance and high fruit quality requires rigorous selection for multiple traits and multiple generation cycles, which for many tree fruit can take decades of breeding, as is the case for the development of fire blight–resistant pears. Long-term testing of pear selections (A) in multiple environments is generally necessary before high quality, fire blight–resistant cultivars such as 'Potomac' (B) and 'Blake's Pride' (C) can be released. Gene transfer (transformation) is being used to improve existing cultivars and to introduce into the germplasm of a species new and useful characteristics such as high-level plum pox virus resistance in C5 plum (D) that resulted from transformation with the plum pox virus coat protein gene. (*Source:* Photos courtesy of Richard L. Bell and Ralph Scorza, U.S. Department of Agriculture, Kearneysville, WV.)

Commercial cherry packing line equipped with two ten-lane parallel zone sizers (insert) and automatic bagging machines (background). Packing capacity is 30 metric tons per hour. (*Source:* Photo courtesy of Stemilt Growers, Wenatchee, WA.)

'Goldrush', a disease-resistant cultivar from a cooperative breeding program of the Agricultural Experiment Stations of Indiana, New Jersey, and Illinois. Through a regional project joining over 50 scientists in the United States and Canada, promising new cultivars are evaluated for insect and disease susceptibility, horticultural characteristics, and organoleptic qualities over a wide range of climatic conditions. (*Source:* Photo courtesy of Stephen S. Miller, U.S. Department of Agriculture, Kearneysville, WV.)

1. NEMATODES
2. NUTRITIONAL VALUE OF FRUIT

Nematodes

John M. Halbrendt

Nematodes are roundworms that can be found in virtually every ecological niche. They have diverse life cycles, including being parasites of plants and animals and living freely in the soil. Plant-parasitic and free-living nematodes are microscopic, and hundreds or even thousands may be present in a few cubic centimeters of soil. Their small size and ability to adapt to severe and changing environments have made nematodes one of the most abundant animal life forms on earth.

BENEFICIAL FREE-LIVING NEMATODES

Nematodes that live in the soil and feed on bacteria and fungi are referred to as "free-living." These nematodes are beneficial and play an important role in recycling plant nutrients. The bacteria and fungi that decompose organic matter utilize the released nutrients to grow and multiply, thus making them unavailable to plants. Free-living nematodes feed on these microbes and release nutrients back to the soil in their excretory products. Free-living nematodes are considered essential components of healthy soil ecosystems.

PLANT-PARASITIC NEMATODES AND ASSOCIATED PROBLEMS

Plant-parasitic nematodes feed on root tips and feeder roots and are capable of causing severe damage. Nematode-infected roots are less efficient at supplying a plant with water and nutrients and thus increase susceptibility to stress factors such as heat, drought, and nutritional deficiencies. Although nematodes may cause yield losses by themselves, they also combine with other soilborne agents, such as viruses, fungi, and bacteria, to cause complex disease situations. In pome and stone fruit, plant-parasitic nematodes reduce tree vigor, predispose trees to diseases, cause replant problems, and transmit viruses.

Vigor and Yield Reduction

Plant-parasitic nematodes limit the availability of nutrients for tree growth and fruit production. Furthermore, trees must expend metabolic activity to repair damage caused by nematode feeding. This is a chronic problem that continues throughout the season. Without control measures, nematode populations often continue to increase. Nematodes become a problem when the population level surpasses the damage threshold; at this point a tree yields less than its potential. The threshold for damage will vary according to a number of factors, including tree variety, age, size, nutritional status, and moisture stress.

Predisposition to Diseases

Research has shown that nematode feeding can produce physiological changes that predispose trees to other problems. For example, feeding by the ring nematode has been linked with susceptibility to bacterial canker and reduced winter hardiness. The basis for predisposition is likely the result of weakened natural defenses against disease and possibly a disruption of the normal hormonal balance in the root system.

Replant Problems

Plant-parasitic nematodes have repeatedly been associated with the poor establishment of new trees on old orchard sites. Evidence suggests that these replant problems are caused by synergistic interactions between plant-parasitic nematodes and other soilborne microflora.

Virus Transmission

Several important virus diseases are transmitted to fruit trees by dagger nematodes. The nematode acquires the virus when it feeds on an infected plant and transmits it when it feeds on healthy tree roots. Because the nematode is an efficient vector, the threshold for damage is much lower than for nematodes that only parasitize the root system.

MAJOR NEMATODES THAT AFFECT FRUIT TREES

Lesion nematodes (*Pratylenchus* species) are very destructive and leave a trail of dead cells in their path as they feed. These wounds are good sites for secondary infections by fungi and bacteria. Young trees are particularly affected, which helps explain the role of nematodes in replant problems. Fruit trees that experience replant problems include peaches, apples, pears, and cherries. Lesion nematodes have a broad host range and high reproductive potential.

Root-knot nematodes (*Meloidogyne* species) cause large galls to develop on tree roots (Figure N1.1). A gall forms when a juvenile nematode initiates a feeding site or "giant cell" near the vascular tissues. The cells surrounding the nematode proliferate, and the root swells. The nematode remains at the feeding site throughout its development. After reaching maturity, a single female may release 200 to 500 eggs into the soil. Heavy infections of root-knot on peach cause stunting, loss of yield, and eventual decline.

Ring nematodes (*Criconemella* species) are relatively easy to identify under the microscope due to thick annulations on their bodies. These nematodes are sluggish and reproduce rapidly. The feeding by ring nematodes can be very damaging, and the problem is especially serious on young trees. These nematodes have been implicated as an important component of the peach tree short life (PTSL) problem. Trees experiencing PTSL usually appear healthy one year and then suddenly collapse and die in early spring the following year. Predisposition to cold injury and bacterial canker are also part of the PTSL complex.

Dagger nematodes (*Xiphinema* species) are not very damaging to tree roots but are serious threats to fruit production because they are

FIGURE N1.1. Peach seedling root system heavily infected with root-knot nematodes (*Source:* Courtesy of Staci Willhide, Pennsylvania State University, Biglerville, PA.)

efficient vectors of certain tree fruit nepoviruses. Viruses transmitted by dagger nematodes include tomato ringspot virus (TmRSV), cherry rasp leaf virus (CRLV), peach rosette mosaic virus (PRMV), and strawberry latent ringspot virus (SLRV).

Several additional plant-parasitic nematodes have routinely been associated with declining orchards, but proof that these nematodes are causal agents of fruit tree disease is lacking and further research is needed. The nematodes most commonly reported include lance nematodes (*Hoplolaimus* species), spiral nematodes (*Helicotylencus* species), and species of *Tylenchorhynchus* and *Cacopaurus*.

NEPOVIRUS DISEASES

A discussion of diseases caused by nepoviruses can be confusing because the same or closely related strains of a virus can cause different diseases in different hosts. For example, stone fruit diseases including *Prunus* stem pitting (PSP), prune brown line (PBL), Stanley constriction and decline (SCAD), and yellow bud mosaic virus (YBMV) are all caused by strains of TmRSV. The same virus also causes apple union necrosis and decline (AUND) (Figure N1.2). Similarly, CRLV causes cherry rasp leaf disease in cherry and flat apple disease in apple. Strawberry latent ringspot virus is a problem in apricot, but the disease is more severe if the tree is simultaneously infected with other viruses, such as necrotic ringspot virus (NRSV). Some nepoviruses, such as TmRSV, are lethal, while others cause decline and reduce the yield and quality of fruit until a fruit tree is worthless. The tomato ringspot virus is perhaps the most widespread and economically important nepovirus on stone fruit throughout North and South America.

DIAGNOSIS OF NEMATODE PROBLEMS

Nematodes usually do not cause distinctive diagnostic symptoms on crops. A sound diagnosis of nematode problems is based on examination of above- and belowground symptoms, field histories, and laboratory analyses of soil and/or plant tissues. A determination of nematode population levels can only be accomplished with a nematode assay. The soil must be representative of the site to be checked, and great care should go into sample collection. Since the results of the assay are affected by the condition of the nematodes, it is also

FIGURE N1.2. Apple tree broken at graft union due to tomato ringspot virus infection (*Source:* Courtesy of Staci Willhide, Pennsylvania State University, Biglerville, PA.)

very important that samples be handled properly until they are delivered to the lab.

NEMATODE CONTROL

Nematode problems in established orchards are difficult to control, and, therefore, good management should focus on preventive measures. As with many other pest problems, nematode control begins with sanitation and good cultural practices. Transplanting of infected stock and movement of infested soil by humans are primary means of dissemination. Interregional and intercontinental movement of infested soil and plant material has almost certainly extended the geographic range of many nematodes. Cultural practices such as optimization of soil pH, proper fertilization, improvement of soil tilth and organic matter, and weed control promote healthy trees and thus minimize the effects of nematode damage.

Some nematode and nepovirus problems can be controlled with genetically resistant rootstocks. When available, this is one of the most economical and environmentally sound methods for managing these problems. However, for most nematode and nepovirus diseases, genetic resistance is not available.

The best time to begin a nematode control program is before a new orchard is planted (Figure N1.3). Soil fumigants are the most efficient and effective chemicals for nematode control, and these can be applied only prior to planting. In general, fumigants are broad-spectrum biocides and also provide some level of disease and weed control; however, efficacy will vary depending upon the particular fumigant used and soil conditions at the time of incorporation. The most volatile soil fumigants require tarping.

Contact nematicides can be used as preplant or postplant treatments. When used as preplant treatments, they can be thoroughly incorporated before the orchard is planted, making broadcast application more effective. These products can be incorporated into the soil with a rototiller or other mechanical means or by irrigation. Nematicides are highly effective, but the level of control is usually not as good as fumigation due to limited movement of the chemicals through the soil. For control of nepovirus diseases, it is also important to effectively manage broadleaf weeds that serve as reservoirs of the virus.

FIGURE N1.3. Nectarine orchard with dead and declining trees due to prunus stem pitting disease, caused by tomato ringspot virus. Replant sites require extensive renovation for successful tree establishment.

The use of a crop rotation for nematode control can provide several benefits. Aside from suppressing nematode populations, rotation crops can also reduce weed problems, increase soil organic matter, improve nutrient availability, and help control erosion. In addition, decomposition of rotation crops improves soil drainage and aeration. This improves tree growth and promotes nutrient recycling. Some crops, such as rapeseed and marigold, also release nematicidal compounds upon decomposition.

Plant-parasitic nematodes are economically important pathogens of fruit trees. Heavily infested orchards may never reach their full production potential and, in the most severe cases, may fail completely due to poor yield, reduced quality, and/or tree death. A preplant soil assay can determine the potential for nematode problems and whether control measures are needed. Preplant detection and treatment of orchard nematode problems is the most economical, efficient, and effective control strategy.

Related Topics: DISEASES; ORCHARD PLANNING AND SITE PREPARATION; PLANT-PEST RELATIONSHIPS AND THE ORCHARD ECOSYSTEM; ROOTSTOCK SELECTION

SELECTED BIBLIOGRAPHY

Nyzcepir, A. P. and J. M. Halbrendt (1993). Nematode parasites of fruit trees. In Evans, K., D. Trudgill, and J. Webster (eds.), *Plant parasitic nematodes in temperate agriculture* (pp. 381-425). Wallingford, UK: CAB International.

Whitehead, A. G. (1998). *Plant nematode control.* Wallingford, UK: CAB International.

Yadava, U. L. and S. L. Doud (1980). The short life and replant problems of deciduous fruit trees. *Hort. Rev.* 2:1-116.

2

Nutritional Value of Fruit

Andrea T. Borchers
Dianne A. Hyson

Nutrient compositions of temperate tree fruit are presented in Tables N2.1 and N2.2. These data are based primarily on analyses provided by the U.S. Department of Agriculture (USDA, 2001). The values represent average composition; it is noteworthy, however, that cultivar, strain, country of origin, maturity and ripeness, and growing, harvesting, storage, and processing conditions can significantly affect fruit nutritional value. In general, however, the USDA values are consistent with other published data.

Potassium is the only mineral listed (Table N2.2), since most temperate fruit contain only minor amounts of other minerals and trace elements such as calcium, iron, phosphorus, and sodium. Almonds are a notable exception, providing 248 milligrams calcium, 4.3 milligrams iron, 275 milligrams magnesium, and 474 milligrams phosphorus per 100 grams of edible portion. The calcium and magnesium contents of figs (35 milligrams and 15 to 212 milligrams per 100 grams edible portion, respectively) are also considerably higher than levels in the other fruit.

HEALTH BENEFITS OF FRUIT CONSUMPTION

Current dietary guidelines recommend the inclusion in the daily diet of several servings of fruit due to their relatively low caloric value and negligible sodium, cholesterol, and fat (with the exception of almonds, which provide approximately 80 percent of energy as fat) (USDA, 2000). More important, the variety and combination of nutrients in fruit and vegetables are thought to have potential health

TABLE N2.1. Nutritional compositions of temperate tree fruit (in mg/100 g edible portion)

Fruit	Latin name	Water	Energy (kcal)	Protein	Total lipid	Carbohydrate	Total dietary fiber	Ash
Almond	*Prunus amygdalus*	5.3	578	21.3	50.6	19.7	11.8	3.11
Apple	*Malus domestica*	84.0	59	0.2	0.4	15.3	2.7	0.26
Apricot	*Prunus armeniaca*	86.4	48	1.4	0.4	11.1	2.4	0.75
Cherry, sour	*Prunus cerasus*	86.1	50	1.0	0.3	12.2	1.6	0.40
Cherry, sweet	*Prunus avium*	80.8	72	1.2	1.0	16.6	2.3	0.53
Fig	*Ficus carica*	79.1	74	0.8	0.3	19.2	3.3	0.66
Mulberry	*Morus nigra*	87.9	43	1.4	0.4	9.8	1.7	0.69
Nectarine	*Prunus persica*	86.3	49	0.9	0.5	11.8	1.6	0.54
Papaw	*Asimina triloba*	(89.3)	(29)	(0.4)	(0.1)	(6.9)	(2.3)	n.a.*
Peach	*Prunus persica*	87.7	43	0.7	0.1	11.1	2.0	0.46
Pear	*Pyrus communis*	83.8	59	0.4	0.4	15.1	2.4	0.28
Persimmon	*Diospyros kaki*	80.3	70	0.6	0.2	18.6	3.6	0.33
Plum	*Prunus domestica*	85.2	55	0.8	0.6	13.0	1.5	0.39
Quince	*Cydonia oblonga*	83.8	57	0.4	0.1	15.3	1.9	0.40

Primary source: USDA reference tables (USDA, 2001). *Source for values in parentheses:* Wills (compiler), 1987, *Food Technol. Austral.* 39:523-526. *n.a. = not available.

TABLE N2.2. Select minerals, vitamins, and phenols in temperate tree fruit (in mg/100 g edible portion unless otherwise indicated)

	MINERALS	VITAMINS				Total Phenols	PHENOLS		
	Potassium	Ascorbic Acid	Toco-pherol	Vitamin A (RE)‡	Caroten-oids		Flavanols	Flavonols	Anthocyanins
Almond	728	0	26.18	1			detected	detected	
Apple	115	5.7 / **7.0-17.8**	0.32	5		229-350 [10] / 110-600 [15, cider apples]	19.3-43.0 [10] / 15.3-56.8 [4, pulp] / 67-232.9 / 9.8-42.9 [4]	47.4-122.0 [4, peel] / 2.64-7.39 [4] / 230 [puree]**	11.8-32.4 [10] / n.d.-104† [4]
Apricot	296	10	0.89	261	6.23 / 0.5-14.0 [9]		0.5 / 4.7-108.9* [9]	9.5-43.8* [9]	detected
Cherry, sour	173	10	0.13	128			12.03***		7.52-23.59 / ~90 mg/100 ml juice
Cherry, sweet	224	7	0.13	21					3.1-14.6 / 82-297 [dark cherries] / 2-41 [light cherries]
Fig	232	2	0.89	14		1090-1110	0.15		
Mulberry	194	36.4 / **10**	0.45	2.5	0.01				1-10
Nectarine	212	5.4	0.89	74	0.06	70-150			
Papaw	**140**	**60**			**0.91**				
Peach	197	6.6	0.7	54	0.15	19.6-29.0 [4] / 29-125 [10] / 46.7-80.1 [8] / 68.8-92.0	8.62 / 5.36-18.7 [8]	0-1.2 [8]	0-1.78 [8]
Pear	125	4	0.5	2	0.01	41.4-56.1 [4]	2.1-2.17 [2] / 18.5-60.2 / 1.5-7.1	21-123.9 [5, peel, none detected in pulp] / 13.0 [puree]*	
Persimmon	161	7.5	0.59	217 / **75-77**	1.5-2.2	3970	1.27		
Plum	172	9.5	0.6	32	0.12	26.2-92.2	49.65		14-20 / 177 [skin]

TABLE N2.2 (continued)

Quince	197	15	0.55	4	0.05	not established	5.24 16.4-306 [commercial jams]	170.3 [puree]** 0-2.98 [commercial jams]
Recommended intake for adults over 18	2400 mg/d	75-90 mg/d	15 mg/d	700-900 µg/day‡	further research needed	not established	not established	not established

Sources: Vitamin and mineral data are mostly from USDA reference tables (USDA, 2001); values in bold are from other literature sources and are provided if no USDA data are available or if considerably different values were noted. Phenol data are from various sources (see the following source list). When available, numbers of cultivars analyzed are shown in brackets. *Sources in addition to USDA*: **Almond**—Almond Board of California, preliminary data. **Apple**—Price et al., 1999, *Food Chem.* 66:489-494; de Pascual-Teresa et al., 2000, *J. Agric. Food Chem.* 48:5331-5337; Podsedek et al., 2000, *Eur. Food Res. Technol.* 210:268-272; Escarpa and González, 2001, *J. Chromatogr. A.* 823:331-337. **Apricot**—Radi et al., 1997, *HortSci.* 32:1087-1091; de Rigal et al., 2000, *J. Sci. Food Agric.* 80:763-768; de Pascual-Teresa et al., 2000, *J. Agric. Food Chem.* 48:5331-5337. **Cherry**—Wang et al., 1997, *J. Agric. Food Chem.* 45:2556-2560; Wang et al., 1999, *J. Agric. Food Chem.* 47:840-844; Petersen and Poll, 1999, *Eur. Food Res. Technol.* 209:251-256; Heinonen et al., 1998, *J. Agric. Food Chem.* 46:4107-4112; Gardiner et al., 1993, *New Zealand J. Crop Hort. Sci.* 21:213-218; Gao and Mazza, 1995, *J. Agric. Food Chem.* 43:343-346. **Fig**—de Pascual-Teresa et al., 2000, *J. Agric. Food Chem.* 48:5331-5337. **Mulberry**—Gerasopoulos and Stavroulakis, 1997, *J. Sci. Food Agric.* 73:261-264. **Peach**—Karakurt et al., 2000, *J. Sci. Food Agric.* 80:1841-1847; Chang et al., 2000, *Agric. Food Chem.* 48:5331-5337; Carbonaro and Mattera, 2001, *Food Chem.* 72:419-424. **Pear**—de Pascual-Teresa et al., 2000, *J. Agric. Food Chem.* 48:5331-5337; Escarpa and González, 2001, *Eur. Food Res. Technol.* 212:439-444; Carbonaro and Mattera, 2001, *Food Chem.* 72:419-424. **Persimmon**—Herrmann, 1996, *Industrielle Obst- und Gemüseverwertung* 81:114-121; Wesche-Ebeling et al., 1996, *Food Chem.* 57:399-403; de Ancos et al., 2000, *J. Agric. Food Chem.* 48:3542-3548; de Pascual-Teresa et al., 2000, *J. Agric. Food Chem.* 48:5331-5337; Bibi et al., 2001, *Nahrung/Food* 45:82-86. **Plum**—Siddiq et al., 1994, *J. Food Process. Preserv.* 18:75-84; Wesche-Ebeling et al., 1996, *Food Chem.* 57:399-403; Herrmann, 1996, *Industrielle Obst- und Gemüseverwertung* 81:114-121; de Pascual-Teresa et al., 2000, *J. Agric. Food Chem.* 48:5331-5337. **Quince**—de Pascual-Teresa et al., 2000, *J. Agric. Food Chem.* 48:5331-5337; Silva et al., 2000, *J. Agric. Food Chem.* 48:2853-2857; Andrade et al., 1998, *J. Agric. Food Chem.* 46:968-972.

*Per 100 g dry matter.

** Purees were obtained by boiling fruit for 15 minutes, then removing cores and pulping them.

*** Not clear whether sweet or sour cherries.

† Measurable amounts of anthocyanidins are present in only red-peeled apples such as 'Delicious'.

‡ RE are still reported for USDA data, although RAE (retinol activity equivalents) are now the preferred method of reporting by the National Research Council. For preformed vitamin A 1RAE = 1RE = 1µg retinol. For conversions between RAE and provitamin A source, refer to Trumbo et al., 2001, *JADA* 101:294-301.

benefits. Numerous epidemiological and some intervention studies indicate that increased consumption of fruit, nuts, and vegetables is associated with decreased risk of heart disease, cancer, and possibly other chronic diseases (Kris-Etherton et al., 1999; Ness and Powles, 1997; Steinmetz and Potter, 1996). Some of the potentially beneficial nutrients found in temperate tree fruit include dietary fiber, potassium, vitamin A and carotenoids, vitamin C, tocopherol, and phenolic compounds.

- *Dietary fiber (both soluble and insoluble forms):* Dietary fiber is important for gastrointestinal health. Studies indicate that it may be associated with reduced risk of certain types of cancer as well as improved control of blood lipids and glucose (Anderson and Hanna, 1999).
- *Potassium:* Many, but not all, cross-sectional and epidemiological studies identify an inverse relationship between blood pressure and the amount of potassium in the diet (Burgess et al., 1999). Potassium may also protect against risk of stroke.
- *Vitamin A and carotenoids:* Vitamin A is required for normal vision and immune function. In addition, consumption of foods rich in preformed and provitamin A may reduce risk of some types of cancer (Lampe, 1999).
- *Vitamin C (ascorbic acid):* Vitamin C provides important antioxidant protection and appears to affect most aspects of the immune system (Hughes, 1999). Diets with high vitamin C content from fruit and vegetables are associated with reduced risk of some types of cancer (Levine et al., 1996).
- *Tocopherol:* Most fruit do not provide significant quantities of tocopherol; however, almonds (as well as all other nuts) are a very good source. Diets rich in food sources of tocopherol are associated with reduced risk of coronary heart disease (Lampe, 1999).
- *Phenolic compounds:* Phenolic compounds are present in all land-based plants, but their distribution is largely genera and species specific. They play a major role in determining taste, flavor, and color. Recent research has focused on the potential health benefits of phenolic compounds, particularly flavonoids. In vitro, phenolic compounds are powerful antioxidants, can modulate platelet activation and endothelial function, and can influence a variety of enzyme activities—all important pro-

cesses in the prevention of cardiovascular disease, cancer, and possibly other chronic diseases. Dietary phenolics are at least partially absorbed and appear to retain some activity in vivo despite extensive metabolic modification and degradation. Evidence from a growing number of animal studies as well as some clinical trials suggests that specific phenolic compounds can lower the risk of some chronic diseases, particularly heart disease and some cancers. Because of their potential importance, the phenolic compounds found in pome and stone fruit are highlighted in the following section.

PHENOLIC COMPOUNDS IN POME AND STONE FRUIT

Quantitative data on total phenols and the major types of flavonoids, the most important class of polyphenols, in pome and stone fruit are incomplete. Furthermore, there is considerable variability in the existing data, even for fruit that have been quite extensively analyzed. Since the flavonoid composition of apples has been investigated in more detail than that of other tree fruit, apples are used in Table N2.3 to illustrate that much of the inherent variability in phenolic content arises from genetic factors. The phenolic content and composition are also greatly influenced by an array of orchard and postharvest factors as well as differences in analytical methodology. The data in Table N2.3 further illustrate the varied distribution of phenolics within the fruit itself. The peel of apples (and most pome and stone fruit) generally has a much higher concentration than the pulp.

In a majority of the pome and stone fruit analyzed to date, flavonols, flavan-3-ols, and anthocyanidins—three subclasses of flavonoids—have been detected. In addition, other types of flavonoids have been reported for specific fruit, e.g., chalcones in apples, flavanones in almonds and tart cherries, and isoflavones in tart cherries. No quantitative data are yet available on these flavanones (naringenin and naringenin-glycosides) in cherries and almonds, but the amounts present in almonds appear to be substantial (Almond Board of California, preliminary data).

Temperate tree fruit are a moderate source of dietary fiber and potassium, and a fair source of vitamin C and vitamin A. Many of the pome and stone fruit contain significant amounts of phenolic com-

TABLE N2.3. Variation in content of select individual phenolic compounds in some apple cultivars

Cultivar/Tissue	Flavan-3-ols	Procyanidin B1	Procyanidin B2	Procyanidin B3	Catechin	Epicatechin
Golden Delicious/cortex					1.4	8.8
Golden Delicious/pulp	15.3	1.0-1.1	2.3-3.2	2.1-2.7	2.8-4.9	1.9-3.4
Golden Delicious/peel	61.7	3.2-5.3	6.9-16.6	2.5-6.6	6.6-16.4	8.2-16.8
Golden Delicious	9.8	0.5	3.2	0.1	0.2	2.0
Granny Smith/pulp	56.8	6.2-8.4	9.7-10.5	6.1-10.0	13.6-18.2	7.1-9.7
Granny Smith/peel	173.7	17.3-24.1	55.8-57.4	7.0-12.4	37.4-48.6	24.6-31.2
Granny Smith	20.0	1.7	4.1	0.2	0.6	2.7
Reinette/pulp	44.9	5.7-6.7	8.2-9.4	3.3-4.1	11.3-13.6	9.1-11.1
Reinette/peel	188.0	10.3-24.2	38.8-58.1	12.5-15.8	22.9-46.0	23.8-43.9
Reinette	42.9	3.7	11.5	0.3	1.4	6.9
Delicious/pulp		1.1-2.1	3.4-5.4	2.0-2.8	4.4-7.0	3.6-5.9
Delicious/peel		12.7-17.2	43.3-65.9	1.1-1.4	29.7-44.5	24.8-48.1
Delicious	38.4	3.4	7.9	0.4	1.6	6.4

Sources: de Pascual-Teresa et al., 2000, *J. Agric. Food Chem.* 48:5331-5337; Escarpa and González, 2001, *Eur. Food Res. Technol.* 212:439-444; Sanoner et al., 1999, *J. Agric. Food Chem.* 47:4847-4853.

pounds, suggesting a potential role of these fruit in promoting human health.

Related Topics: FRUIT COLOR DEVELOPMENT; POSTHARVEST FRUIT PHYSIOLOGY

SELECTED BIBLIOGRAPHY

Anderson, J. W. and T. J. Hanna (1999). Impact of nondigestible carbohydrates on serum lipoproteins and risk for cardiovascular disease. *J. Nutr.* 129:1457S-1466S.

Burgess, E., R. Lewanczuk, P. Bolli, A. Chockalingam, H. Cutler, G. Taylor, and P. Hamet (1999). Recommendations on potassium, magnesium and calcium, *CMAJ* 160:S35-S45.

Hughes, D. A. (1999). Effects of dietary antioxidants on the immune function of middle-aged adults. *Proc. Nutr. Soc.* 58:79-84.

Kris-Etherton, P. M., S. Yu-Poth, J. Sabate, H. E. Ratcliffe, G. Zhao, and T. D. Etherton (1999). Nuts and their bioactive constituents: Effects on serum lipids and other factors that affect disease risk. *Amer. J. Clin. Nutr.* 70:504S-511S.

Lampe, J. (1999). Health effects of vegetables and fruits: Assessing mechanisms of action in human experimental studies. *Amer. J. Clin. Nutr.* 70:475S-490S.

Levine, M., S. Rumsey, Y. Wang, J. Park, O. Kwon, W. Xu, and N. Amano (1996). Vitamin C. In Ziegler, E. E. and L. J. Filer (eds.), *Present knowledge in nutrition* Seventh edition (pp. 146-159). Washington, DC: ILSI Press.

Ness, A. R. and J. W. Powles (1997). Fruit and vegetables, and cardiovascular disease: A review. *Int. J. Epidemiol.* 26:1-13.

Steinmetz, C. A. and J. D. Potter (1996). Vegetables, fruit, and cancer prevention: A review. *J. Amer. Diet. Assoc.* 96:1027-1039.

U.S. Department of Agriculture, Agricultural Research Service (2001). *USDA nutrient database for standard reference,* Release 14. Retrieved October 1, 2001, from <http://www.nal.usda.gov/fnic/foodcomp>.

U.S. Department of Agriculture, Health and Human Services (2000). *Nutrition and your health: Dietary guidelines for Americans,* Fifth edition, Home and garden bull. 232. Washington, DC: USDA.

1. ORCHARD FLOOR MANAGEMENT
2. ORCHARD PLANNING AND SITE
PREPARATION

– 1 –

Orchard Floor Management

Ian A. Merwin

Orchards are unique among crop systems in their temporal and structural complexity. During the 15- to 50-year production cycles of perennial fruit plantings, a diverse community of naturally growing "weeds" or planted groundcover species develops on the orchard floor. This groundcover vegetation can provide substantial benefits of soil conservation, nutrient cycling, and habitat for desirable wildlife. However, without careful management it can also compete with trees for limiting nutrients, complicate orchard operations, and harbor economic pests of fruit. Sustainable orchard floor management (OFM) systems require knowledge about site-specific conditions, plant function, and consideration of trade-offs among beneficial and detrimental aspects of groundcover vegetation.

ORCHARD GROUNDCOVER ADVANTAGES

Soil fertility provides the foundation for productivity in any crop system, and it is especially important in perennial crop systems where there are few options for supplementing nutrient availability in the deeper root zone, because direct placement of fertilizer into the rhizosphere is difficult and can damage roots. Furthermore, stringent soil, climate, and infrastructure requirements of orchards usually cause growers to replant the same or similar fruit crops repeatedly in the same locations over many decades or centuries. Long-term maintenance of soil fertility and structure is especially important because serious soil problems can develop over time. Orchard soils are prone to wind or water erosion; compaction by tractors, sprayers, and harvest operations; and gradual increases in soilborne pathogens that infect tree roots. Groundcovers provide a renewable surface layer of biomass that protects soil from weathering and compaction and influences populations of beneficial and detrimental soil microorganisms. As this biomass decomposes into the mineral soil, it replenishes organic matter—promoting microbial activity, sustaining soil nutrient reserves, and increasing soil pore volume and water-holding capacity (Hogue and Neilsen, 1987).

Deciduous fruit trees in cool-climate regions remain dormant for almost half of the year, and there is little uptake of essential nutrients from soil by dormant trees. The potential for soil erosion and leaching or runoff of nitrogen, phosphorus, and pesticide residues is greatest during the dormant season. Cool-season grasses such as *Festuca* and *Lolium* species, and broadleaf groundcovers such as brassicas and legumes that continue growing when fruit trees are dormant, can serve as "green manure" or "relay" cover crops that fix or retain nitrogen and other essential nutrients in biomass residues during the winter months. Mowing or tillage of these groundcovers in late spring releases nutrient reserves at a time when they are readily assimilated by fruit trees. Growing dormant season groundcovers that are tilled or killed with herbicides the following spring is increasingly popular in fruit-growing regions with mild winters where soil and nutrient conservation are high priorities (Marsh, Daly, and McCarthy, 1996; Tagliavani et al., 1996).

Groundcovers growing between or within the tree rows can also be managed to help control tree vigor, enhancing fruit quality and tree winter hardiness. Excess soil nitrogen and water availability during late summer and early autumn can prolong shoot and canopy growth, delay fruit maturation, increase the potential for winter cold injury when woody tissues fail to harden-off sufficiently, and make dormant season pruning more difficult and costly (Elmore, Merwin, and Cudney, 1997; Merwin and Stiles, 1998). Encouraging moderate groundcover growth and competition in early autumn can be accomplished either by seeding fast-growing cover crops in late summer or by timing nonresidual herbicide applications or cultivation so that groundcovers naturally reestablish from seeds or root propagules at the desired time of year. The advantages of managing, rather than eliminating, groundcover competition for water and nutrients during critical times of year have been widely recognized by the wine grape industry, where market incentives for fruit qualities essential to make high-value wines offset the losses in gross yield that may result from late-season groundcover competition with vines (Elmore, Merwin, and Cudney, 1997). The tree fruit industry could benefit from similar efforts to develop OFM practices that control canopy vigor in order to improve eating quality and consumption of fresh fruit.

Permanent grass and broadleaf groundcovers also facilitate access by orchard customers, workers, and machinery during wet/muddy or dry/dusty conditions. Pick-your-own growers need to consider subjective factors that encourage patrons to visit their farms, and a well-mowed green sward is generally more attractive to the public than the bare soil of "weed-free" plantings. In orchards where dropped fruit are gathered for processing or fermentation, it is especially important to minimize mud splashing and soiling of fruit beneath trees, and grass or clover groundcovers are often maintained over the entire orchard floor.

Certain groundcover species can suppress pest insects, fungi, and nematodes that parasitize fruit trees. For example, preplant cover crops of marigold *(Tagetes patula)* or cereal wheat *(Triticum aestivum)* can suppress pathogenic nematodes and fungi (respectively) that damage apple roots and cause soilborne "replant disease." Flowering groundcovers that provide habitat, pollen, and nectar food sources for predatory insects, such as hover flies (Syrphidae) and assassin bugs (Reduviidae), can increase populations of these beneficial insects in orchards and help to control leaf-feeding pests such as aphids and caterpillars, reducing the need for pesticides. Selecting and utilizing specific groundcovers to promote beneficial insects, fungi, and soil microbial activity in orchards is an integrated pest management (IPM) tactic that merits renewed attention from researchers (Brown and Glenn, 1999).

Orchard groundcovers of various types and mixtures are important components in sustainable fruit-growing systems. Properly managed groundcovers can improve soil fertility and water-holding capacity, facilitate access during inclement weather, provide habitat for beneficial wildlife, make orchards more attractive for workers and pick-your-own customers, suppress pathogenic soil fungi and nematodes, and limit excess vigor of mature bearing trees—optimizing fruit quality and reducing pruning costs.

ORCHARD GROUNDCOVER DISADVANTAGES

Despite the potential advantages of groundcovers, in most fruit plantings the surface vegetation is suppressed or eliminated over all or part of the orchard floor. In some situations this is a matter of practical convenience. For example, in almond groves where the nuts are gathered by vacuuming or sweeping from the ground surface, harvest is more efficient when no groundcover residues are present. Where flood irrigation is practiced, horizontal flow of irrigation water is more rapid and smooth when the orchard floor is completely weed free and uniform, although infiltration of irrigation or rainwater into most soils is enhanced when surface structure and porosity have been protected by groundcovers (Elmore, Merwin, and Cudney, 1997). In most situations, the primary reason for suppressing or eradicating orchard groundcovers is that, without proper management, they become "weeds" that compete with the crop for limiting nutrients and water, reducing tree growth and productivity.

Fruit trees have relatively sparse root systems that do not compete effectively with most groundcovers for water and nutrients. Grass and herbaceous groundcover root systems are more dense and pervasive, and excavation studies show that there are few tree roots in the topsoil beneath orchard groundcovers, compared with weed-free herbicide-treated areas in tree rows (Atkinson, 1980). Isotope tracer studies reveal that nitrogen fertilizers applied to the orchard floor are almost completely assimilated by soil microbes and groundcover vegetation, with little short-term uptake by trees. Soil water content during midsummer is also reduced beneath vegetative groundcovers in comparison with weed-free tree rows, despite greater tree root density in weed-free soil, which indicates that fruit trees use water more sparingly than groundcovers. Heavy irrigation is usually necessary to provide sufficient water for trees that must compete with groundcovers for soil nutrients—even in regions with humid growing seasons. Frequent mowing reduces only slightly the evapotranspiration of soil water by groundcover vegetation, although studies demonstrate a negative correlation between groundcover biomass per square meter and soil water availability during the growing season (Elmore, Merwin, and Cudney, 1997; Merwin and Stiles, 1998). In regions where rodents, rabbits, or hares often cause serious damage to the trunks and lower branches of fruit trees, OFM systems that reduce the height and density of groundcovers help to limit depredation by these pests.

Weedy groundcovers can be especially damaging in young orchards where trees must rapidly establish root and shoot systems that fill their allocated space in the orchard. The optimal type and proportional area of groundcovers in orchards thus depend upon numerous site-specific factors. Determining acceptable levels of groundcover competition—the economic damage threshold at which soil conservation and other pest management benefits compensate for acceptable levels of groundcover competitive interactions with the crop—is complex and variable from region to region, depending upon orchard planting systems, climates and soil types, other management practices or constraints, and marketing strategies (Elmore, Merwin, and Cudney, 1997).

The most common OFM systems in deciduous orchards involve permanent groundcover mixtures of perennial grasses and herbaceous broadleaf species, such as clovers, maintained by periodic mowing in the drive alleys between tree rows, and strips beneath the trees that are treated with herbicides, mulches, or cultivation to suppress groundcovers during the growing season or year-round. The relative widths of the drive lane groundcover and tree row weed suppression strips can be adjusted as appropriate for tree age and vigor or soil nutrient and water supply. In numerous studies comparing different ratios of groundcover to weed-free area, the optimal weed-free area varies according to tree age and soil conditions (Hogue and Neilsen, 1987). In high-fertility soils where tree roots are concentrated by drip irrigation and weed suppression, studies show that fruit production is equivalent in weed-free strips of 2, 4, and 6 square meters per tree (Merwin and Stiles, 1998). Under nonirrigated conditions or in soils with low nutrient reserves, growth and productivity of young trees generally increase as the weed-free area increases, up to 10 square meters per tree, beyond which there is little further gain. Less is known about the importance of weed-suppression timing in orchards, but a few studies suggest that controlling weed competition during the early months of the growing season is especially important for successful establishment of young fruit trees.

Herbicides, mulches, or mechanical cultivation can all provide sufficient control of groundcover competition within tree rows, but these three OFM practices differ substantially in cost, convenience, and soil impacts (Elmore, Merwin, and Cudney, 1997; Hogue and Neilsen, 1987). Preemergence residual or postemergence herbicides are the easiest and least expensive methods for suppressing weeds. Preemergence soil-active herbicides can be safely applied beneath established trees in most situations, and a single application can control most groundcover species for the entire growing season or longer. However, the residual soil persistence and activity of these herbicides can exacerbate the risk of chemical leaching or transport of eroded soil particles and prolong selective pressure for weed genotypes with herbicide resistance. It may not be desirable or necessary to keep tree rows weed free during the dormant season when soil weathering and nutrient losses are most likely (Merwin and Stiles, 1998; Tagliavanni et al., 1996).

Postemergence nonresidual herbicides suppress groundcovers for a relatively brief time (usually four to eight weeks) during the growing season and permit surface vegetation to reestablish later in autumn, thereby providing soil organic matter and surface protection during the dormant season. Long-term OFM studies indicate that tree growth and productivity are as good or better when sparse groundcovers are allowed to reestablish in tree rows treated with nonresidual herbicides earlier in the dormant season, compared with soil maintained completely weed free year-round with residual herbicides (Marsh, Daly, and McCarthy, 1996; Merwin and Stiles, 1998). Considering that soil porosity, nutrient and water retention, and organic matter content are better conserved in the sparse seasonal groundcover that results from nonresidual postemergence herbicide treatments, the use of residual herbicides to maintain the orchard floor continuously weed free arguably constitutes "overkill" in many situations.

Cultivation with rototillers, disks, and harrows or moldboard plowing to mechanically suppress weeds was a common practice for several centuries, and it is still common in orchards with deep, coarse-textured soils. Over the long term, mechanical cultivation often depletes soil organic matter, degrades soil structure, increases erosion and nutrient runoff, and ultimately fails to control weeds that regenerate from rhizomes and other root propagules. Cultivation within crop rows also destroys part of a tree's upper root system, which can be a serious problem for young trees and in shallow soils. Specialized orchard cultivation equipment has been developed to minimize some of these problems, and periodic tillage is still a useful OFM system when used in conjunction with dormant season cover crops in regions with mild winters and soils that are not prone to erosion.

Synthetic fabrics, or "geotextiles," plastic films, and various natural biomass mulches provide nonchemical alternatives for suppressing groundcover vegetation in tree rows. They can serve to protect roots and soil from deep-freezing in the winter and to raise soil temperatures earlier in the spring (in the case of black plastic or fabrics). Mulches are rather expensive to apply and may require supplemental hand weeding to control certain species. They can also increase damage to trees by voles (*Microtus* species), and most of the synthetic fabric or film mulches are not biodegradable. Biomass mulches increase organic matter and supplement nitrogen, phosphorus, potassium, calcium, magnesium, and other essential nutrients in the soil— traits that make these mulches especially useful on coarse-textured, droughty soils with low water-holding capacity and fertility (Merwin and Stiles, 1998).

So-called "living mulches" of clovers, vetches, and other legumes have been evaluated as sources of nitrogen and weed suppression for orchards. In regions where legume groundcovers can be grown in the dormant season and suppressed or tilled into the soil at the start of the growing season, they are a useful tactic for weed control and soil conservation. In colder regions where living mulches complete growth and development during summer months, they are usually too competitive for water and other limiting nutrients to provide much benefit to fruit trees.

Orchard floor management is an important part of a fruit-growing system. Groundcover vegetation has both detrimental *and* beneficial roles in orchards. Effective OFM involves recognizing and understanding the positive and negative interactions between fruit trees and groundcovers within a complex agroecosystem. No single OFM program is optimal for all situations, but various effective strategies do provide a satisfactory balance between groundcovers as a tactic for IPM and soil conservation and groundcovers as problematic weeds. Short-term gains in fruit production and management convenience must be evaluated with due regard for long-term priorities such as soil and water quality and the sustainability of fruit-growing systems.

Related Topics: NEMATODES; ORCHARD PLANNING AND SITE PREPA-
RATION; PLANT-PEST RELATIONSHIPS AND THE ORCHARD ECOSYS-
TEM; SOIL MANAGEMENT AND PLANT FERTILIZATION; SUSTAINABLE
ORCHARDING; WATER RELATIONS; WILDLIFE

SELECTED BIBLIOGRAPHY

Atkinson, D. (1980). The distribution and effectiveness of the roots of tree crops. *Hort. Rev.* 2:424-490.

Brown, M. and D. M. Glenn (1999). Ground cover plants and selective insecticides as pest management tools in apple orchards. *J. Econ. Entomol.* 92:899-905.

Elmore, C. L., I. Merwin, and D. Cudney (1997). Weed management in tree fruit, nuts, citrus and vine crops. In McGiffen, M. E. (ed.), *Weed management in horticultural crops* (pp. 17-29). Alexandria, VA: ASHS Press.

Hogue, E. J. and G. H. Neilsen (1987). Orchard floor vegetation management. *Hort. Rev.* 9:377-430.

Marsh, K. B., M. J. Daly, and T. P. McCarthy (1996). The effect of understory management on soil fertility, tree nutrition, fruit production and apple fruit quality. *Biological Agric. and Hort.* 13:161-173.

Merwin, I. A. and W. C. Stiles (1998). *Integrated weed and soil management in fruit plantings,* Info. bull. 242. Ithaca, NY: Cornell Coop. Ext. Serv.

Tagliavani, M., D. Scudellazi, B. Marangoni, and M. Toselli (1996). Nitrogen fertilization management in orchards to reconcile productivity and environmental aspects. *Fert. Research* 43:93-102.

Orchard Planning and Site Preparation

Tara Auxt Baugher

Mistakes made in planning and planting an orchard are difficult to reverse. Before establishing a new orchard block, astute growers carefully assess all the factors that will ultimately affect fruit quality, production efficiency, and orchard sustainability. Proper planning includes evaluations of business goals, management style, site characteristics, global planting trends, regional production statistics for different systems, and market potential. University extension services offer enterprise budgets for calculating internal rate of return or net present value for various fruit crops and systems. Fruit growing is a high-risk venture, and many new computer programs allow growers to conduct sensitivity analyses. Optimal site preparation and planting involve thinking in terms of managing tree roots for increased orchard performance. Physical, chemical, and biological properties of the soil must be considered. Soil structure is a major concern on a new site. A replant site requires extensive renovation to avoid tree mortality or stunted growth.

SITE ASSESSMENT

Cold-air drainage and soil quality have significant effects on the profitability of an orchard. An ideal site is on the upper side of a gradual (4 to 8 percent) slope, on rolling or elevated land. Low-lying areas, where cold air can accumulate during a calm, clear night, are prone to spring frost damage. Hilltops or ridges may expose trees to excessive winds or to arctic air masses. A preferred orchard soil is a deep (at least 1 meter), well-drained, and aerated loam. Detailed soil appraisals should be conducted several years in advance of planting.

A grower begins by obtaining a soil map and by digging test holes to examine the soil profile. Soil maps provide useful information on soil texture, parent material, native fertility, erosion levels, and water-holding capacity. Test holes reveal impervious layers and water-related problems. If checked several times during a rainy period, the pits will yield valuable information on the soil water table. Topsoil and subsoil samples also are collected at this time for analysis of pH, nutrient deficiencies and toxicities, and organic matter content. Separate samples are collected for evaluation of replant disease factors, such as nematodes and herbicide residues. Additional site considerations include access to water for irrigation and spraying, the presence of weeds that serve as reservoirs for plant viruses, and the potential for hail or other weather-related disasters.

ORCHARD DESIGN AND TREE QUALITY

Substantial thought should be given to orchard design. Important considerations are canopy light interception and distribution to flowers and fruit. Research in temperate regions shows that trees grown in north-south oriented rows have better light conditions than those grown in east-west rows. Decreasing the distance between rows and increasing tree height also increases light interception. With most tree forms, optimum tree height is half the row spacing plus 1 meter. Maximizing production per hectare by planting trees in high densities requires careful assessment of the vigor potential of a site. It is helpful to evaluate tree size in a previous orchard or in an adjacent block. Other factors that affect decisions on tree arrangement include topography, equipment size, and worker access.

To obtain the scion/rootstock combinations best suited to an orchard plan, growers order trees two to three years ahead of planting. Ordering virus-tested trees with a strong root system ensures a good start for a sustainable production system. Well-feathered trees are desirable for early cropping, intensive systems. Windbreak trees, if needed, and pollinizer trees also should be ordered early. Studies indicate that the best trees for windbreaks are alders *(Alnus)*, willows *(Salix)*, or other deciduous species that leaf out early in the spring and hold leaves past harvest time. Fruit tree bloom periods vary from one region to another, and it is wise to get local advice on pollinizers.

ORCHARD PREPARATION

The one chance a grower has to optimize the soil environment is prior to planting. Before disturbing the surface vegetation, spot treatments can be made to control perennial and other problem weeds. On replant sites, a cover-cropping system can be established and maintained for several years to suppress weeds, nematodes, and soilborne fungi and to increase soil organic matter. Disinfecting soils is another approach to improving early growth and yield on old orchard sites. Soil drainage problems should be corrected with subsurface drainage systems or surface modifications such as ridging. Stone fruit and certain dwarf apple rootstocks are especially sensitive to waterlogging and associated diseases caused by *Phytophthora* species. Some form of deep soil manipulation should be employed to break up fragipans and to loosen and mix horizons. For best results, the soil should be friable. After deep chiseling or subsoiling in four directions, the site is replowed to incorporate lime and fertilizer. The chemical status of the soil is ameliorated to the depth of the root zone, since lime and phosphorus are not very mobile and potassium moves slowly.

After the soil is thoroughly prepared, an orchard groundcover is established. Turf grasses often are the most desirable groundcovers, especially species that suppress voles, broadleaf weeds, and soilborne problems. Grasses also conserve nutrients, increase organic matter, protect groundwater quality, and improve water infiltration. To prevent erosion, the groundcover should be established shortly after the site is cultivated and leveled. Grass seed can be sown in the row middles, leaving 1.5- to 2.5-meter-wide bare strips where the trees are planted, or seed can be sown over the entire field. In the latter system, the sod is established at least one season before planting and later killed, leaving a mulch that enhances early tree growth.

TREE PLANTING

Several studies show that time of planting greatly affects initial tree growth. Early planted trees have increased shoot numbers and length, and fewer trees become spur-bound or stunted. Orchards should be planted as early in the spring as the ground can be worked or in late fall in regions where sudden drops in temperature are unlikely. Mechanical tree planters, developed in the 1970s, make it possible to complete planting in a shorter time frame, when soil moisture conditions are optimal. Research on tree planting indicates that trees may be planted by a variety of methods, provided close root-soil contact is secured and the trees are not planted too deeply. With a tree planter, it is important to adjust tree height by hand and to tamp the soil firmly around the roots. If an auger is used to drill the planting holes, the glazed edges of the hole must be fractured to permit root penetration. To prevent scion rooting, the bud union of dwarf trees should be 5 centimeters above the soil line. Higher bud union placement is generally avoided due to the potential for burr knots or winter injury on some rootstocks. Soon after planting, the trees should be watered and, if needed, a support system established.

The goal of advance planning and site preparation is to ensure early and regular crops of high-value fruit for the 15- to 30-year life of an orchard. Preplant use of sustainable management practices guarantees that a site will support the current orchard and generations to come.

Related Topics: CULTIVAR SELECTION; GEOGRAPHIC CONSIDERATIONS; IRRIGATION; MARKETING; NEMATODES; ORCHARD FLOOR MANAGEMENT; ROOTSTOCK SELECTION; SOIL MANAGEMENT AND PLANT FERTILIZATION; SPRING FROST CONTROL; SUSTAINABLE ORCHARDING

SELECTED BIBLIOGRAPHY

Autio, Wesley R., Duane W. Greene, Daniel R. Cooley, and James R. Schupp (1991). Improving the growth of newly planted apple trees. *HortScience* 26:840-843.

Auxt, Tara, Steven Blizzard, and Kendall Elliott (1980). Comparison of apple planting methods. *J. Amer. Soc. Hort. Sci.* 105:468-472.

Baugher, Tara Auxt and Rabindar N. Singh (1989). Evaluation of four soil amendments in ameliorating toxic conditions in three orchard subsoils. *Applied Agric. Res.* 4:111-117.

Biggs, Alan R., Tara Auxt Baugher, Alan R. Collins, Henry W. Hogmire, James B. Kotcon, D. Michael Glenn, Alan J. Sexstone, and Ross E. Byers (1997). Growth of apple trees, nitrate mobility and pest populations following a corn versus fescue crop rotation. *Amer. J. Alt. Agric.* 12:162-172.

Fuller, Keith (2000). Bolstering the soil environment: Site preparation. *Compact Fruit Tree* 33:25-27.

1. PACKING
2. PHYSIOLOGICAL DISORDERS
3. PLANT GROWTH REGULATION
4. PLANT HORMONES
5. PLANT NUTRITION
6. PLANT-PEST RELATIONSHIPS
AND THE ORCHARD ECOSYSTEM
7. POSTHARVEST FRUIT PHYSIOLOGY
8. PROCESSING
9. PROPAGATION

Packing

A. Nathan Reed

Fresh fruit packers and storage operators endeavor to maintain fruit freshness and deliver attractive fruit of uniform color and weight, free from external blemishes and insect damage. Their ultimate goal is to furnish fruit that are not only aesthetically appealing but also free from internal defects and possessing outstanding flavor, aroma, and texture properties. Since fruit are living, biological products that are grown in many locations, they are affected by many factors, including weather, soil conditions, horticultural practices, insect and fungus populations, and postharvest storage conditions. Packing and marketing perfect fruit is an ideal situation that is difficult to achieve.

PRESIZING AND PRESORTING

Depending on the potential marketing window, the storability, the bruising susceptibility, and how perishable the crop is, some fruit are put through an initial process of presizing. Presizing is a sorting process that accumulates fruit of similar sizes and colors for future packing. The durability of some apple cultivars and their ability to be stored for long periods in a controlled atmosphere make them the most likely candidates for presizing. The packing operation can run much more efficiently if large, uniform lots of fruit are being processed together. Without presizing, the potential number of commercial package combinations of sizes and grades possible from a single group of fruit could number as high as 150. Presizing is also a filtering mechanism that can eliminate culls due to insect, pathogen, or mechanical injury. Removal of culls thus improves the efficiency of the storage and packing system by decreasing the energy consumed in storage and/or handling and materials and consumables used in packing. Most, if not all, cultivars of stone fruit are so fragile and perishable that presizing is not a viable option. Packing stone fruit therefore is an intensive operation that must be carried out immediately following harvest and must account for many variables or have marketing tolerances of greater variability in color and size within the same box.

WEIGHT SIZING, COLOR SORTING, AND PACKAGING

Apples, Pears, Peaches, Nectarines, Plums, and Apricots

Several manufacturers offer graders or sizers with high-speed carrier systems capable of handling 12 to15 fruit per second and accurately segregating them based on weight with an accuracy of +/– 1 gram (Figure P1.1). These systems are used frequently in handling and sorting individual apples, pears, apricots, nectarines, and peaches. Some cultivars of pears with irregular shapes or elongated necks

FIGURE P1.1. Commercial packing line with 'Golden Delicious' apples prior to color grading and weight sizing (*Source:* Photo courtesy of Stemilt Growers, Wenatchee, WA.)

make handling more difficult. Most of this equipment has camera options available for capturing color images of the entire surface of each fruit. Using computer control systems, sorting decisions can be made based on programmable criteria and the amount and shades of color present on each fruit. Fruit color standards (and also quality and packing grade standards) have been established by the U.S. Department of Agriculture and other government and private agencies (http://www.ams.usda.gov/standards/frutmrkt.htm). In some situations, nonmechanized packing occurs directly out of picking containers into trays, and the packer performs all sorting and sizing functions manually.

Hand packing machine-sorted fruit into shipping and retail display cartons has long been a strenuous and labor-intensive operation. Loose filling boxes can be accomplished mechanically, but fruit must be bruise resistant and tolerant of the impacts that occur in this method. Recently, robotic equipment has been developed that handles this delicate operation at commercially viable speeds approaching 1.5 boxes per minute.

Fruit-packing orientation and materials are very important. Fruit that look good going into the shipping carton may not look good upon arrival at their destination because the wrong packing materials were used. During shipping, fruit are exposed to vibrations and changes in acceleration that can result in significant bruising. Compression bruising can also appear if cartons collapse or fruit are packed too tightly. If fruit are not immobilized in the carton, they can develop scuffing marks from the friction of vibration. All these factors can result in the rejection of the product by retailers and/or consumers. It is imperative to select packing materials that protect the fruit, minimize weight loss, are lightweight to reduce shipping and energy costs, and are recyclable and environmentally friendly.

Cherries

Optical sorting equipment, developed in Walla Walla, Washington (http://www.keyww.com/), for use in the food-processing industry, has been adapted and used in sorting cherries based on color. This equipment relies on multiple cameras positioned to observe individual cherries that are distributed into individual lanes on a vibratory conveyor and then launched into the air for surface color evaluation. The computer processors are extremely quick to recognize cherries "in-flight," capture the surface color information, and then make a determination as to whether fruit meet specific criteria. The decision to reject an individual fruit based on its lack or abundance of a color is made, and the cherry is blown out of its trajectory by a series of individually controlled air jet nozzles. The nozzles deliver enough air pressure to divert individually selected fruit to a separate water flume for transport to a separate packing area. This equipment is very sophisticated and is capable of processing up to 14 metric tons of cherries per hour.

Sizers in use for smaller stone fruit such as cherries are not as prevalent as those used in apples and pears. Weight sizing for cherries is impractical due to time restrictions of singulating individual fruit and the large volume of fruit being processed by commercial packing operations. Cherry sizers have been developed to sort fruit based on fruit dimensions instead of weight. However, cherries also create special problems with respect to size, since they are not spherical or symmetrical and are harvested with stems attached. The first cherry sizers were developed using inclined, diverging rollers. Gravity and water assist cherries as they traverse downward along the sloping, spinning rollers. Smaller fruit exit the rollers near the top of the incline while the largest fruit travel to the largest gap at the lower end of the rollers. Four or five water flumes under the rollers carry the "sized" cherries to box-filling equipment.

The latest technology in cherry sizing is a system developed in Wenatchee, Washington (http://www.stemilt.com/). This technology also relies on stainless steel rollers that rotate at over 300 rotations per minute, but the rollers in this version are parallel. The number of parallel zones and their gap width determine the number of sizes to which cherries can be sorted. The cherry sizer is computer controlled and capable of changing roller gaps to the nearest 0.1 millimeter. Each sizing lane is capable of sizing more than 1.5 metric tons of cherries per hour.

The fact that this equipment is highly specialized and expensive is exaggerated by the reality that cherries are quite perishable and have a very narrow harvest window. The entire packing season in the northwestern United States is spread over about eight weeks due to differences in maturity timing associated with cultivar and orchard elevation. Without the high market value of cherries, this technology would not be economically feasible.

DEFECT SORTING

Another strenuous and labor-intensive operation is sorting for external defects. A few manufacturers offer automated equipment to do this task. It is a challenging task to electronically emulate the human vision and mental decision processes, and to guarantee 100 percent inspection, identification, and classification of each individual fruit. Today there is no perfect or error-free computer vision system available in the marketplace.

MODIFIED-ATMOSPHERE PACKING

Modified-atmosphere packing (MAP), also known as "maintain and preserve," is a technology employed to maintain and preserve the quality of fruit being packed and shipped. Modifying the atmosphere in which a fruit is stored can dramatically alter its rate of respiration and metabolism and increase its commercial storage life. A passive or active process can generate modified atmospheres. In the passive mode, respiring fruit, over time, generate their own atmosphere of elevated carbon dioxide and reduced oxygen. In an active system, desired endpoint concentrations are established more quickly by flushing the fruit with a desired mixture of gases.

Successful MAP depends on several factors. Each packed container has a bag or film material, usually a type of polyethylene, that surrounds the product and provides a barrier to gas transfer between the inside and outside. The oxygen transmission rate of the film will limit what products can be stored within each container. Other limiting factors include the respiration rate of the product, the amount of product, the temperature at which the fruit are being stored, and the amount of free space within the film.

Temperature is the major factor to manage in all postharvest handling. In MAP, temperature is very critical and must be maintained within specific limits. As temperature increases, so does the respiration rate. This is a dangerous situation for fresh fruit in a sealed bag and might result in the consumption of all available oxygen and the onset of anaerobic respiration, the production of carbon dioxide, ethanol, and off flavor.

A MAP alternative for fruit with high respiration rates utilizes microperforation or combines the polyethylene film with a more porous membrane material attached as a patch. Microperforated film allows greater gas exchange and is more forgiving during episodes of moderate temperature abuse. Microperforations in films are produced by mechanical methods of laser, sharp-pointed instruments, or localized heat. The porous patch acts as a valve and allows greater exchange of gases than does the polyethylene film by itself.

A California corporation (http://www.landec.com/) has developed an "intelligent" polymer technology that is useful in MAP situations where temperature abuse is likely or insurance against improper temperature handling is warranted. The temperature switch technology uses a polymer that changes from a crystalline to a more fluid form in response to increasing temperature. As the polymer becomes more fluid, it allows greater exchange of gas and thus compensates for increased respiration rate by allowing more oxygen to diffuse into the package. The polymer changes its permeability over the range of a few degrees. Once recooled, the process reverses, the polymer acts as a barrier, and the fruit attains an equilibrium of oxygen and carbon dioxide relative to its respiration rate and temperature. The permeability of the polymer, or the "switch," can be adjusted to match critical temperatures and respiration rate requirements of individual products.

BRUISE PREVENTION

Bruising is an issue that must be addressed throughout the entire postharvest handling process. Bruising can result from impacts with blunt or sharp objects, or it can result from compression in storage or shipping containers. Bruising at each step of the postharvest handling system can result in significant economic losses. There are several low-technology solutions for bruising problems. Most involve altering the process of how fruit are handled by eliminating drops between transport or fruit carriers; eliminating edges and sharp corners on packing equipment; and the use of fiber-filled or bubble pads in packing cartons. Bruises might not be apparent immediately, but with time they become quite obvious. Finding the origin of the problem can be difficult. A few commercial companies have built instrumented spheres, which are devices that can be placed directly into the product stream.

The instrumented spheres are exposed to the same forces and impacts that fruit experience during the handling and packing operation. The instrumented spheres are high-technology devices with onboard accelerometers that electronically record changes in acceleration along with a corresponding time stamp. By videotaping the sphere as it progresses through the packing line and making comparisons to the time-stamped acceleration data, one can pinpoint where handling problems occur. Based on this information, corrections can be made, tested, and evaluated.

Storage of the product also has a significant effect on how fruit respond to the packing process itself. With 'Golden Delicious' and other light-colored cultivars that easily show bruising, it is a common practice to store fruit at 10 to 15°C for a few days prior to packing. During this time, fruit dehydrate and become more elastic and less susceptible to bruising during packing.

NONDESTRUCTIVE QUALITY ASSESSMENT

Today, the majority of internal quality assessments for lots of fruit are based on destructive tests on selected samples. From a relatively small number of samples, decisions are made about the remaining population. Retail customer requirements for specific internal quality conditions have become increasingly difficult to achieve. The invisible quality attributes that are being promoted and requested include firmness, texture, sweetness, flavor, and assurance of no internal decay or disorders. A number of technologies are being explored in the quest for methods to inspect individual fruit and pack only those which meet minimum criteria and customer expectations.

A nondestructive online method to replace the destructive penetrometer method of measuring firmness is being sought. A possible solution for fruit of moderate firmness, e.g., stone fruit, is being developed in the United Kingdom (http://www.sinclair-intl.com/). This nondestructive method uses an accelerometer in the form of a small bullet probe that taps the fruit and monitors the change in slope of acceleration of the probe during impact. Once developed, this technology can be quickly integrated in conjunction with the equipment that applies labels on individual fruit. For firmer fruit such as apples and pears, a company in the Netherlands (http://www.aweta.nl) is developing an acoustical device that taps fruit and measures the frequency at which it vibrates as a result of the nondestructive impact. A combination of the frequency and the weight of the fruit results in a measurement that can be used for sorting. Consumers can distinguish fruit of differing acoustic levels as firm or soft textured.

Near-infrared (NIR) is a technology that has just recently become commercially available for measuring the sugar content of stone and pome fruit (Figure P1.2). With NIR technology, an association is built between the patterns of the absorbed invisible wavelengths and the internal organic sugar molecules present in the flesh of a fruit. Computer modeling plays a large role in these systems. The operation and results of NIR are dependent on fruit temperature. Under stable operating conditions, NIR is accurate to within approximately a 1 percent sugar level. The reflectance method measures the sugar content near the surface of the fruit. Transmission requires more light but measures more tissue.

Magnetic resonance imaging (MRI) is another promising technique that has been explored. This technology is dependent on the influence of a strong magnetic field on hydrogen nuclei. Fruit are com-

FIGURE P1.2. Near-infrared technology measuring sugar content of apples on a commercial packing line (*Source:* Photo courtesy of Stemilt Growers, Wenatchee, WA.)

posed mainly of water (approximately 85 percent) and thus have an abundance of hydrogen atoms that respond to magnetism. MRI is a valuable technique that is commonly used in the medical field. It also works well for identifying internal defects in fruit. The drawbacks with this technique are its relatively slow speed and the significant costs associated with a need for multiple lanes to sort fruit at commercial rates.

Another accepted technique that is used widely in the medical field today is X-ray technology. Research in this area has shown positive results with identifying internal damage caused by tunneling insects. The technique has much potential for commercialization from a cost and speed standpoint. There is concern, however, regarding the perception of the public with respect to consuming fruit that have been irradiated.

The packing operation is a very critical step in the process of providing fruit to the consumer. Packing is unique to all other operations of growing, storing, handling, and distributing fruit. The packing line acts as a funnel of opportunity for selecting fruit with appealing characteristics, both external and internal, and thus providing the consumer with consistent satisfaction that produces repeat purchases.

Related Topics: FRUIT MATURITY; HARVEST; MARKETING; POST-HARVEST FRUIT PHYSIOLOGY; STORING AND HANDLING FRUIT

SELECTED BIBLIOGRAPHY

Abbott, J. A., R. Lu, B. L. Upchurch, and R. Stroshine (1997). New technologies for nondestructive quality evaluation of fruits and vegetables. *Hort. Rev.* 20:1-120.

Baritelle, A. L. and G. M. Hyde (2001). Commodity conditioning to reduce impact bruising. *Postharvest Biology and Technol.* 21:331-339.

Crisosto, C. H., D. Slaughter, D. Garner, and J. Boyd (2001). Stone fruit critical bruising thresholds. *J. Amer. Pomological Soc.* 55:76-81.

Lange, D. L. (2000). New film technologies for horticultural products. *HortTechnol.* 10:487-490.

LaRue, J. H. and R. S. Johnson (1989). *Peaches, plums and nectarines: Growing and handling for fresh market*, Pub. 3331. Davis, CA: Univ. of California Coop. Exten. Serv., Div. of Agric. and Nat. Resources.

Northeast Regional Agricultural Engineering Service (1997). *Sensors for nondestructive testing: Measuring the quality of fresh fruits and vegetables.* Proceedings from the sensors for nondestructive testing international conference. Ithaca, NY: NRAES.

Wills, R., B. McGlasson, D. Graham, and D. Joyce (1998). *Postharvest: An introduction to the physiology and handling of fruit, vegetables and ornamentals,* Fourth edition. Adelaide, South Australia: Hyde Park Press.

– 2 –

Physiological Disorders

Christopher B. Watkins

Physiological disorders of fresh crops occur as a result of altered metabolism in response to imposition of stresses and are manifested as visible symptoms of cellular disorganization and cell death. Most physiological disorders of temperate fruit occur after harvest, when they are removed from supplies of water, nutrients, hormones, and energy from the tree. Therefore, fruit have an altered ability to respond to stresses in the environment that interrupt, restrict, or accelerate normal metabolic processes in an adverse or negative manner. Interestingly, however, many postharvest management regimens beneficially utilize stress conditions such as temperature and atmosphere modification to maximize storage potential of fruit. During the postharvest period, stress is an external factor that will result in undesirable changes *only* if the plant or plant part is exposed to it for a sufficient duration or sufficient intensity, and, therefore, the postharvest period can be seen as a time of stress management (Kays, 1997).

Physiological disorders are distinct from the many other undesirable postharvest changes in quality, such as water loss, softening, loss of chlorophyll, and other ripening-related events associated with normal senescence, which affect storage potential and thus marketability of fruit. The definition also excludes a number of direct postharvest injuries that can occur as a result of mechanical damage (e.g., bruising), freezing injuries, and exposure to gases or chemical solutions (e.g., ammonia leaks in cold storage or salts and antioxidants used in postharvest treatments). Pathological disorders are also distinct, but it is not uncommon for diseases to be associated with physiological disorders, especially as secondary infections.

A common feature of all physiological disorders is that susceptibility to injury is affected greatly by cultivar, through characteristics such as skin diffusivity to oxygen and carbon dioxide, cell wall and membrane properties, mineral composition, and antioxidant status. Preharvest factors, which include climate, maturity at harvest, nutrition, and orchard management methods, also affect these characteristics. Considerable variation in the resistance of a given fruit to imposed stress occurs, and, therefore, severity and timing of disorder expression can be observed among apparently similar lots of the same cultivar or strain.

TYPES OF PHYSIOLOGICAL DISORDERS

Many physiological disorders have been identified in temperate fruit crops, especially the apple. Photographs of most of these are available in sources such as Lidster, Blanpied, and Prange (1999) and Snowdon (1990). Generally, symptoms of physiological disorders are well defined, but understanding of the biochemical processes involved in their development is incomplete.

Physiological disorders can be considered in three categories: those which develop (1) only on the tree, e.g., watercore and sunscald on apples; (2) on the tree and during storage, e.g., bitter pit of apples; and (3) only during storage. The third category includes most physiological disorders (Table P2.1) and is complex because of the many postharvest management techniques used to maintain quality. These disorders can be divided into those associated with senescence, low temperatures, and use of inappropriate atmospheres during storage.

Senescent disorders are related to harvest of overmature fruit and/ or fruit with nutritional imbalances such as high nitrogen and low calcium contents. Storage at higher than optimal temperatures can also contribute to disorder development.

Chilling injuries are visual manifestations of cellular dysfunction in crops exposed to chilling temperatures. There is some controversy as to whether certain low-temperature disorders occurring in temperate crops are manifestations of chilling injury, but the physiological and biochemical mechanisms of injury probably are identical in all susceptible commodities, and only the rate at which these changes occur differs. Temperature–exposure time interactions exist in development of chilling injury, and dysfunctions induced by chilling temperatures can be repaired upon transfer of the crop to nonchilling temperatures before permanent injury has occurred. The primary re-

TABLE P2.1. Postharvest physiological disorders of apples, pears, peaches, nectarines, plums, and cherries and major effecting factors

Effecting Factors	Apples	Pears	Peaches and Nectarines	Plums	Cherries
Climate, fruit maturity, nutrition	Senescent breakdown, bitter pit, lenticel blotch, Jonathan spot, lenticel spot	Bitter pit, corky spot, cork spot, Anjou pit, breakdown, senescent scald, vascular, internal, and core breakdowns		Internal breakdown	
Storage temperature, fruit maturity, climate	Superficial scald, low-temperature breakdown, soft scald, brown core	Superficial scald	Woolliness, internal breakdown		Surface pitting
Storage atmosphere	Low-oxygen injury, epidermal cracking, ribbon scald, brown heart, external carbon dioxide injury	Brown core, pithy brown core, flesh browning	Low-oxygen injury		Low-oxygen injury, high–carbon dioxide injury

221

sponse to direct chilling stress is thought to be physical in nature, centered on the cell membranes (Figure P2.1). The secondary events include the multitude of metabolic processes that are adversely affected as a consequence of the primary event and lead to visible symptoms and cell death. The subdivision between primary and secondary events is not arbitrary, as it is proposed that it allows the time-dependent secondary events ("effects") to be conceptually separated from the more instantaneous primary event ("cause").

Disorders associated with low oxygen and high carbon dioxide occur when fruit are subjected to atmospheres outside safe limits at any temperature-time combination. The safe limits can vary by fruit type, cultivar, and strain. Damage may be manifested as irregular ripening, initiation and/or aggravation of certain physiological disorders, development of off flavors, and increased susceptibility to decay. Tolerances of fruit to storage atmospheres are affected by metabolic and physical factors and can vary greatly among species, cultivars and strains, maturity and ripening stages, and growing conditions.

Lowered oxygen and elevated carbon dioxide concentrations affect respiration and associated metabolic pathways of glycolysis, fermentation, the tricarboxylic acid cycle, and the mitochondrial respiratory chain, as well as pathways involved in secondary metabolism such as production of ethylene, pigments, phenolics, and volatiles. Increased carbon flux through the fermentation pathway is a common feature of fresh crops exposed to anaerobic conditions, but direct evi-

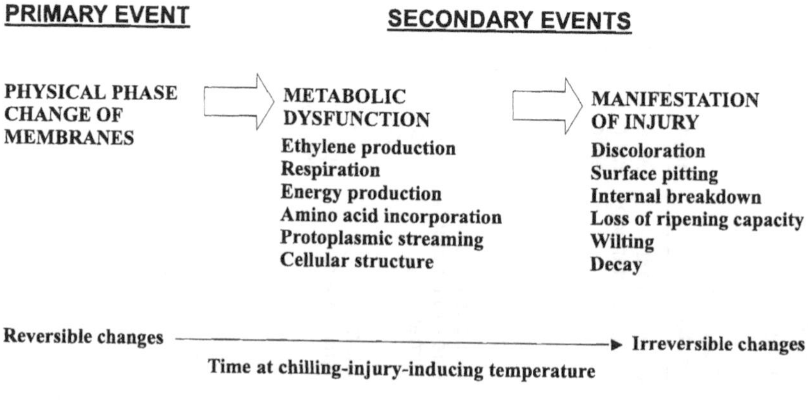

PRIMARY EVENT   **SECONDARY EVENTS**

PHYSICAL PHASE CHANGE OF MEMBRANES	⇒	METABOLIC DYSFUNCTION	⇒	MANIFESTATION OF INJURY
		Ethylene production		Discoloration
		Respiration		Surface pitting
		Energy production		Internal breakdown
		Amino acid incorporation		Loss of ripening capacity
		Protoplasmic streaming		Wilting
		Cellular structure		Decay

Reversible changes ——————————————————————➤ Irreversible changes
Time at chilling-injury-inducing temperature

FIGURE P2.1. A schematic representation of responses of plant tissues to chilling stress (*Source:* Modified from Wang, 1990.)

dence for injury by acetaldehyde and ethanol accumulations has not been demonstrated. Fruit exposed to high carbon dioxide, but usually not low oxygen, show high accumulations of succinate, which may be toxic to plant cells and is thought to be responsible for carbon dioxide injury. However, recent evidence with tissues of different susceptibilities to carbon dioxide injuries has not supported this view (Fernández-Trujllo, Nock, and Watkins, 2001). It is likely that damaging levels of carbon dioxide result from progressive failure to maintain energy balance and metabolic cell function, rather than accumulation of any single injurious compound.

MAJOR PHYSIOLOGICAL DISORDERS OF TEMPERATE TREE FRUIT

The major physiological disorders of apples, pears, peaches and nectarines, and cherries are described in this section, with their causes and control methods. Where disorders are prevalent mainly in apples rather than in pears, e.g., soft scald and watercore, they are described only for apples.

Apples

Watercore

Symptoms are a glassy, water-soaked appearance of the flesh resulting from accumulation of liquid, predominantly sorbitol, in the intercellular spaces (Marlow and Loescher, 1984). Watercored tissues are usually associated with vascular bundles of the core line, although other tissues may be affected. These include the flesh near the surface, which may develop watercore as a result of heat stress and, in severe cases, can be observed through the skin. Watercore develops only on the tree and can lead to crinkle, a disorder characterized by breakdown of the flesh and shallow depressions on the skin. During storage, the disorder can dissipate, but fruit with severe watercore can develop tissue breakdown.

Watercore occurs in most commercial cultivars, but some are more susceptible than others. In 'Fuji', its presence may be desirable, while in other cultivars such as 'Delicious', risk of watercore breakdown during storage has resulted in development of strict grade standards that prevent packing of affected fruit for markets. Development of watercore is associated with harvest of overmature fruit and/or low night temperatures. Watercore development may be due to changes in membrane integrity, rather than inability to metabolize sorbitol. Timing of harvest is the primary method to avoid or obtain (if desired) fruit with watercore.

Sunscald

Sunscalded areas, often golden-brown patches, occur on the exposed cheeks of fruit. The damaged areas may darken in storage. Sunscald development is a nonenzymatic and nonoxidative process. All cultivars can be damaged by sunscald, and because its development cannot be prevented by postharvest treatments such as diphenylamine (DPA), sunscalded fruit should be removed during grading.

Bitter Pit

Symptoms are discrete necrotic lesions on the skin and/or in the flesh, often occurring at the calyx end of the fruit first. Bitter pit is predominantly a storage-related disorder, although it is sometimes discernable on the tree. Another bitter pit-related disorder is lenticel blotch. Bitter pit is cultivar specific, with 'Cox's Orange Pippin' and 'Cortland' being particularly susceptible. Susceptibility to bitter pit is associated with early harvest and preharvest factors that result in low fruit calcium, such as low crop load and large fruit.

Both pre- and postharvest methods may be used to control risk of bitter pit development (Ferguson and Watkins, 1989). Preharvest methods include management practices to improve calcium availability in the soil, such as lime application, pruning and thinning practices that reduce competition between fruit and leaves, and application of calcium salt sprays during the growing season. Postharvest methods include harvesting more mature fruit, drenching fruit with calcium salts, rapid cooling of fruit, and application of storage conditions that delay fruit ripening, e.g., controlled atmosphere (CA). Prediction techniques for bitter pit risk, based on calcium and other minerals, have been developed in some growing regions.

Jonathan Spot and Lenticel Spot

In these disorders, brown to black spots develop on the skin, particularly on the blushed side of the fruit. Jonathan spot is characterized
by haloes surrounding the lesions that occur randomly over the fruit
surface, while lenticel spots may be slightly depressed areas around
the lenticels. The disorders are reduced by avoiding excess nitrogen,
harvest at optimum maturity, rapid cooling, and keeping fruit at the
optimum storage temperature.

Senescent Breakdown

Senescent breakdown occurs widely in fruit stored at higher than
optimal temperatures or for too long. Flesh softening is followed by
development of mealiness and browning. The skin and flesh may
split in advanced cases. Other senescent breakdown–like disorders
that are recognized in the literature are McIntosh breakdown, Jonathan breakdown, and Spartan breakdown.

Breakdown incidence can be reduced by pre- and postharvest calcium applications to fruit, harvest at the optimum stage of maturity,
prompt cooling, and storage at optimum temperatures and humidities.
CA storage generally reduces senescent breakdown incidence.

Superficial Scald (Storage Scald)

Development of superficial scald is associated with long-term, low-
temperature storage of apples and is probably a chilling injury. Susceptibility to scald is affected by cultivar, growing region, and harvest date.
'Delicious' and 'Granny Smith' are highly susceptible, while 'Gala',
'Empire', and 'Braeburn' are scald resistant. Cooler growing regions
have a lower scald risk, apparently related to the cooler nights that are
experienced by the fruit before harvest. Typically, more mature fruit
have lower scald risk than those harvested earlier.

Superficial scald is controlled by postharvest drenches of DPA.
Product labels regulate the maximum DPA concentrations that should
be used on specific cultivars to ensure control with a minimal risk of
chemical damage. Risk of chemical injury increases if DPA is not
discarded when soil accumulates in the solution, or if DPA is used
with chlorine. DPA residues are not allowed on fruit by some import-

ing countries, and there is concern about possible consumer issues with postharvest chemical use. Therefore, nonchemical methods of control for superficial scald, such as low-oxygen CA storage, are used in some growing regions. A fungicide to reduce decay incidence, and calcium salts to reduce bitter pit or senescent breakdown incidence, are often applied with DPA.

Low-Temperature Breakdown

Symptoms include a diffuse browning of the outer cortex that develops into breakdown. Low-temperature breakdown can be distinguished from senescent breakdown by the occurrence of a band of unaffected tissue under the skin, dark vascular strands, and moistness of the tissue. However, at advanced stages it can be difficult to separate the two disorders.

Susceptible cultivars include 'Bramley's Seedling', 'Cox's Orange Pippin', 'Empire', 'McIntosh', and 'Jonathan', when stored at temperatures less than 2 to 3°C. Disorder incidence increases with prolonged storage. Preharvest factors include low crop loads, large fruit, and cool weather during the latter part of the growing season. Low calcium and phosphorus concentrations in the fruit may be associated with higher susceptibility. High humidity and elevated carbon dioxide concentrations in the storage can increase fruit susceptibility to the disorder.

The primary control method is to maintain higher storage temperatures for susceptible cultivars. Because fruit maintain better firmness at lower temperatures, susceptible cultivars are sometimes kept at potentially injurious temperatures for short periods, the length of which varies by cultivar and growing region. Stepwise lowering of temperatures during the storage period has been utilized, but may detrimentally affect fruit quality (Little and Holmes, 2000). Fruit are typically stored at slightly higher storage temperatures under low-oxygen CA storage.

Soft Scald (Deep Scald)

Symptoms are discrete brown lesions that are smooth and slightly sunken where the underlying tissue has become affected. The flesh tissue is initially pale brown, soft, spongy, and moist and is sharply demarcated from the unaffected tissue. A similar disorder, known as ribbon scald, occurs as a low-oxygen injury.

Soft scald is a low-temperature injury of certain cultivars kept at less than 3°C. Susceptible cultivars include 'Jonathan' and 'Honeycrisp'. Susceptibility to soft scald is greater in fruit harvested later than earlier, which may be related to higher fruit respiration rates when cooled. In general, delays between harvest and storage increase injury development. Orchard factors that increase fruit susceptibility to the disorder are dull, cool, wet summers; light crops; large fruit; and vigorous trees on heavy soils. Control methods rely mainly on harvesting fruit at a less mature stage and use of storage temperatures above 2 to 3°C.

Brown Core (Coreflush)

This disorder is known as brown core in North America and coreflush elsewhere. Its symptoms include a pinkish or brownish discoloration of the core tissue, either as a diffuse circular area or as individual angular areas between the seed cavities. The discolored flesh tends to be firm and moist. Susceptible cultivars include 'Granny Smith' and 'McIntosh'.

Brown core incidence may be aggravated by late harvest and cool growing seasons and is more common in cooler than warmer growing regions. It is essentially a low-temperature disorder, but incidence has also been related to high carbon dioxide in CA storage and senescence. Control procedures include harvest at optimum maturity, avoiding low storage temperatures, and low-oxygen and low–carbon dioxide CA storage.

Low-Oxygen Injury

Symptoms consist of brownish areas with definite margins on the skin, which can range from small patches to most of the fruit. Internally, brownish corky sections occur, with occasional cavities that may be contiguous with external injury. Additional symptoms are alcoholic off flavors, brownish flesh discoloration caused by alcohol injury, bleaching or scalding of the skin, and purpling of the blushed areas of the skin. Skin purpling may be the first visual symptom of low-oxygen injury but is not always evident. Early stages of off-flavor development and skin darkening can be reversed by removal of fruit to air.

Other forms of low-oxygen injury are epidermal cracking and ribbon scald. In epidermal cracking, the flesh tissue is usually dry and mealy and yields readily to pressure but is distinct from mealy breakdown. It may be aggravated by high humidity in the storage atmosphere. Ribbon scald appears as smooth, brown, irregular-shaped, well-defined lesions of the skin.

Cultivars vary in susceptibility to low-oxygen injury and in expression of injury symptoms, probably as a function of sensitivity of tissues to low oxygen concentrations and to physical features such as skin characteristics that affect gas diffusion into the fruit. Factors that increase risk of low-oxygen injury include late harvest, delays between harvest and application of CA storage, slow fruit cooling, and low storage temperatures. Low-oxygen injury is prevented by maintaining oxygen concentrations above the minimum for the cultivar, and by avoiding the factors described earlier. Other considerations include the length of exposure of fruit to low oxygen, carbon dioxide concentrations, and the storage temperature. Usually, storage temperatures for low-oxygen CA storage are higher than for standard CA storage.

Brown Heart (Core or Flesh Browning)

Affected fruit have patches of brown flesh, which may be distributed randomly or as a zone between the core and the flesh, depending on the cultivar. Usually fruit appear externally normal. However, the disorder can be observed on the fruit surface in severe cases. The brown tissue is initially firm and moist but may become dry with cavity formation. Development of brown heart usually ceases when causal conditions are removed.

The primary cause of brown heart is elevated carbon dioxide in the storage atmosphere, damage being related to concentration and length of exposure to the gas. Injury is usually associated with incorrect CA storage but can occur in air storage conditions where ventilation is poor, e.g., in cartons and in ship holds. Apple fruit are less sensitive to elevated carbon dioxide than pear fruit. However, in both fruit types, cultivar effects are important, perhaps reflecting anatomical differences such as size of intercellular spaces and rates of gas diffusion in the tissues.

Other factors affecting susceptibility to brown heart include growing region, orchard, and harvest date. Disorder risk is increased with more mature fruit, large fruit size, delayed cooling, low storage temperature, and low oxygen. The importance of each factor can vary greatly by cultivar. Delays between harvest and exposure to elevated carbon dioxide can reduce susceptibility of fruit to brown heart. In some cultivars, maintenance of low carbon dioxide during the first four to six weeks of CA storage is recommended to minimize risk. DPA used to control superficial scald also controls carbon dioxide injury, and fruit losses have occurred when DPA use has been discontinued.

External Carbon Dioxide Injury

Symptoms of external carbon dioxide injury are irregularly shaped, colorless, brown, or black lesions on the skin, often partly sunken with sharply defined edges. Appearance of injury under elevated carbon dioxide concentrations can occur within a few weeks. Factors that affect fruit sensitivity to the disorder are similar to those described for brown heart, except that early rather than late harvest increases injury risk of susceptible cultivars. Carbon dioxide concentration and duration of exposure, rapid establishment of high carbon dioxide before fruit are cooled, and presence of free moisture on the fruit surface can affect disorder incidence. DPA treatments control external carbon dioxide injury.

Pears

Bitter Pit

This disorder is also known as corky spot on 'Packham's Triumph' (South Africa), and cork spot or Anjou pit in 'D'Anjou' pears (United States). Symptoms may appear both before and after harvest. Bitter pit in pears is similar to that described for apples, and similar control methods apply.

Breakdown

External symptoms include skin yellowing during storage, while internally flesh softening, breakdown, and browning develop. A number of related disorders that develop in pears with breakdown, including senescent scald, vascular breakdown, internal breakdown, and core breakdown, are described by Snowdon (1990).

As with senescent breakdown of apples, senescent breakdown in pears is a disorder of overmature fruit that are cooled slowly and stored at temperatures above the optimum, or for extended periods. The disorder, therefore, can be controlled by attention to these factors.

Superficial Scald (Storage Scald)

Development of superficial scald on pears is associated with long-term storage and has most of the characteristics described for apples. Scald is controlled by using postharvest ethoxyquin drenches, as DPA is not registered for pears.

Low-Oxygen and High–Carbon Dioxide Injuries

Brown core of 'D'Anjou' pears, pithy brown core of 'Bosc', and flesh browning or cavitation of 'Bartlett' occur with certain oxygen and carbon dioxide combinations (Lidster, Blanpied, and Prange, 1999). In brown core and pithy brown core, the core tissue surrounding the carpel turns brown, followed by cavity formation, while in flesh browning the cortex tissues next to the vascular region and outside the core line turn slightly brown and show many little cavities. Dark brown skin discoloration of 'D'Anjou' pears occurs under prolonged CA storage with low oxygen, high carbon dioxide, or a combination of both. These injuries can be avoided by maintaining appropriate storage atmospheres.

Peaches and Nectarines

Woolliness

Symptoms are the development of mealy texture, a lack of flavor and juiciness, and a failure of fruit to ripen when removed from cold storage. Susceptibility to woolliness development is associated with late-maturing cultivars, cool growing seasons, and harvest of less-mature fruit. Disorder incidence can be reduced by cultivar and harvest management and postharvest techniques such as modified atmosphere storage, delayed storage, ethylene treatments, and intermittent warming of fruit during storage.

Internal Breakdown

Symptoms are diffuse, internal discoloration of the flesh and dry, soft flesh. The disorder mostly appears after transfer from low temperatures to ripening temperatures.

Low-Oxygen Injury

Low oxygen results in development of intense skin browning and grayish brown breakdown of the flesh near the skin or surrounding the stones. Low-oxygen injury can be distinguished from internal breakdown by presence of both external and internal symptoms, well-defined areas of injury, browning, and flesh injury that is not necessarily dry (Lidster, Blanpied, and Prange, 1999). Symptoms can appear at any time during storage. Injury can be avoided by maintaining appropriate storage atmospheres.

Plums

Internal Breakdown

Symptoms are internal browning near the stone followed by breakdown of the affected tissue into a gelatinous mass. Development of internal breakdown usually occurs after harvest, but it can be observed before harvest. Orchard and climactic factors, such as hot weather, predispose fruit to development of breakdown.

Cherries

Surface Pitting

Symptoms are irregular depressions that occur on the shoulders or sides of fruit (Looney, Webster, and Kupferman, 1996). Pitting results from impact damage, where cells beneath the skin dehydrate when injured. The majority of pitting occurs during packing operations. Warmer cherries are more resistant to damage and develop fewer pits than colder cherries when subjected to the same forces.

Storage at temperatures near 0°C or the transfer of fruit from cold storage to room temperature worsens pitting.

Control measures include minimizing damage events during harvesting and handling. Low-oxygen, high–carbon dioxide, and high-humidity atmospheres do not affect surface pitting incidence, but preharvest sprays or postharvest dips of calcium salts can decrease its incidence on 'Van' sweet cherries.

Low-Oxygen and High–Carbon Dioxide Injuries

Cherry stems are more sensitive than flesh to high–carbon dioxide (development of red-brown color) and low-oxygen (development of black-brown color) atmospheres (Lidster, Blanpied, and Prange, 1999). Fruit darkening associated with cell membrane rupture and leakage can occur after removal from inappropriate CA conditions. High carbon dioxide can also cause droplets of exudates to form on the fruit, followed by surface browning. These injuries can be avoided by maintaining appropriate storage atmospheres.

Many physiological disorders are known in temperate tree fruit. Most industries have the knowledge about cultivar susceptibility, orchard management, and storage techniques to minimize risk, although the impact of preharvest factors, especially climate, can markedly affect susceptibility and thus potential losses due to these disorders. Even though many postharvest injuries are caused by inappropriate storage regimens, there is major variation among cultivars of fruit in response to these treatments. A closer linkage between plant breeders, who often select material solely based on appearance and productivity, and postharvest scientists to ensure that selection decisions also incorporate susceptibility of fruit to physiological disorders should be encouraged.

Related Topics: FRUIT MATURITY; HARVEST; PACKING; POSTHARVEST FRUIT PHYSIOLOGY; PROCESSING; STORING AND HANDLING FRUIT

SELECTED BIBLIOGRAPHY

Ferguson, I. B. and C. B. Watkins (1989). Bitter pit in apple fruit. *Hort. Rev.* 11:289-355.

Fernández-Trujllo, J. P., J. F. Nock, and C. B. Watkins (2001). Superficial scald, carbon dioxide injury, and changes of fermentation products and organic acids in 'Cortland' and 'Law Rome' apple fruit after high carbon dioxide stress treatment. *J. Amer. Soc. Hort. Sci.* 126: 235-241.

Hansen, E. and W. M. Mellenthin (1979). *Commercial handling and storage practices for winter pears,* Agric. exper. station special report 550. Corvallis, OR: Oregon State Univ.

Kays, S. J. (1997). *Postharvest physiology of perishable plant products.* Athens, GA: Exon Press.

Lidster, P. D., G. D. Blanpied, and R. K. Prange, eds. (1999). *Controlled-atmosphere disorders of commercial fruits and vegetables,* Pub. 1847/E. Ottawa, Ontario: Agric. Agri-Food Canada.

Little, C. R. and R. J. Holmes (2000). *Storage technology for apples and pears.* Knoxfield, Victoria, Australia: Dept. of Nat. Resources and Envir.

Looney, N. E., A. D. Webster, and E. M. Kupferman (1996). Harvesting and handling sweet cherries for the fresh market. In Webster, A. D. and N. E. Looney (eds.), *Cherries: Crop physiology, production and uses* (pp. 411-441). Oxon, UK: CAB International.

Marlow, G. C. and W. H. Loescher (1984). Watercore. *Hort. Rev.* 6:189-251.

Snowdon, A. L. (1990). *A color atlas of post harvest diseases and disorders of fruits and vegetables,* Volume 1. Boca Raton, FL: CRC Press.

Wang, C. Y. (1990). *Chilling injury of horticultural crops.* Boca Raton, FL: CRC Press.

– 3 –

Plant Growth Regulation

Christopher S. Walsh

The regulation of plant growth and form has been a topic of intense biological research. In horticulture, considerable scientific effort has been spent studying hormonal development, with the goal of improving fruit productivity and quality. Much of this work has focused on the exogenous applications of plant growth regulators (PGRs). PGRs are chemicals that mimic hormonal effects on development. In some cases, the hormone itself can be reapplied as a PGR. For example, gibberellin can be applied in the field to reduce russet on apples, and ethylene can be applied in storage to preripen peaches and nectarines. But, in most cases, PGRs are exogenously applied chemicals that are structurally similar to the endogenous plant hormone but are resistant to inactivation in the field. Consequently, PGRs can mimic endogenous hormones but can be synthesized and applied far less expensively than the endogenous plant hormone.

While this discussion is intended to give relevant information to answer production questions, it presents an overview of processes and materials, rather than exact chemical recommendations. Chemical registrations, product names, and formulations are constantly changing. For this discussion to remain useful, it is necessary to present a broader picture of strategies needed, rather than focusing on particular products and rates. Timely information on rates and products is available by consulting commercial labels and local recommendations.

PGR usage in temperate fruit trees can be grouped into four broad categories: (1) regulation of tree vigor and the enhancement of flowering, (2) chemical thinning, (3) control of preharvest drop of fruit, and (4) specialty applications in targeted situations. In these four categories, fruit growers are most likely to use chemical thinning and preharvest drop control annually, if the appropriate materials are registered for their crop. Regulation of vigor, enhancements of flowering, and specialty applications are available but are typically used on a small proportion of fruit acreage, such as young trees, or to address problems inherent in a particular cultivar.

REGULATION OF TREE VIGOR AND ENHANCEMENT OF FLOWERING

In modern orchards, young trees are protected from weeds and pests, fertilized heavily, and irrigated to induce rapid filling of their allotted spaces. Then, in the third or fourth leaf, growers expect trees to decrease vegetative growth and begin flowering and fruiting, so the orchard will be profitable and tree size will be limited. Growers rely on size-controlling rootstocks to switch trees from a high-vigor to a precocious state. This strategy is not always successful. For example, even in the presence of size-controlling stocks, apple cultivars such as 'Jonagold', 'Mutsu', and 'Fuji' may be excessively vigorous for intensive plantings. In other species, such as peach and sweet cherry, size-controlling stocks suitable for North America are still under test and not yet widely used by the industry.

To solve these problems, numerous PGRs have been tested and registered for use in commercial orchards. The underlying hypothesis in vigor control is that there is a "hormonal balance" in the young tree. When that balance favors vigor, flowering, and hence fruiting, is suppressed. When vigor is reduced, flowering and fruiting will follow.

Regulation of vigor in young trees is modulated using PGRs that shift the tree balance from vegetative to reproductive. Direct inhibitors of gibberellin synthesis are compounds such as paclobutrazol and uniconazole. These directly reduce extension growth, induce terminal bud formation, and lead to an increase in flower bud initiation. Once flowering and fruiting begin, cropping controls vigor further, so that a tree can be managed within its allotted space (Zimmerman, 1972). Applications are typically made as a soil drench or spray early in one growing season, with the goal of enhanced yields in the subsequent season.

PGRs that act indirectly to enhance flowering are compounds such as ethephon. Ethephon is applied early in the growing season to nonbearing trees and is metabolized to release ethylene. The ethylene released then suppresses extension growth, induces radial growth and branch stiffening, and enhances flower bud initiation.

Selecting one of the previous approaches for managing young trees depends on crop, PGR cost, and grower preference. Pome fruit growers are likely to use ethephon, while stone fruit growers are more likely to choose gibberellin biosynthesis inhibitors. Controlling tree vigor is more difficult in stone fruit than in pome fruit due to the lack of adequate size-controlling rootstocks; thus, direct inhibition of growth through suppressing gibberellin biosynthesis is required. Additionally, application of ethephon can cause phytotoxicity in stone fruit trees. Characteristic symptoms induced by ethephon are leaf abscission and gummosis.

CHEMICAL THINNING

Chemical thinning is the regulation of crop load through the addition of PGRs that reduce flowering and crop load in established orchards. Used correctly, chemical thinners can have two dramatic effects: (1) increased fruit size and quality and (2) maintenance of annual bearing.

In pome fruit, chemical thinners are commonly applied between full bloom and the cessation of cell division, which occurs about a month after full bloom. The exact timing of thinner application depends on material chosen, cultivar, and local conditions. Table P3.1 lists the most widely used chemical thinners for pome fruit, as well as their benefits and weaknesses.

Chemical thinning of stone fruit has been far less successful than in pome fruit. In pome fruit, chemical thinning takes advantage of the inherent differences in seed count, fruit development, and sink strength to remove smaller-size fruit. Since stone fruit are single seeded and little difference exists among fruitlets, chemical thinning is more difficult. Fewer materials are available for commercial orchards. To circumvent this problem, thinning strategies that rely on prebloom or bloom thinning can be used. The major prebloom thinner available is gibberellic acid. Since stone fruit do not have mixed buds, flower differentiation during the previous summer determines whether a bud will be vegetative or reproductive. For years, tart

TABLE P3.1. A listing of the commonly used plant growth regulators for chemical thinning of pome fruit and some benefits and weaknesses of each

Material	Type of PGR	Benefits	Weaknesses
Carbaryl	Insecticide	Unlikely to overthin, useful on easy-to-thin cultivars	Can affect beneficial insects
Dinitro compounds (DNOC)	Phytotoxic to flowers	Useful for bloom thinning, inexpensive	Environmental concerns, dry weather required
Ethephon	Ethylene-releasing compound	Long period of activity, relatively inexpensive	Temperature sensitivity, premature ripening
Accel™	Cytokinin-plus-gibberellin mixture	Promotion of early season fruit growth	Relatively untested, expensive
Naphthalene acetic acid (NAA)	Synthetic auxin	Inexpensive, long history of usage	Pygmy fruit, temporary growth suppression
Naphthalene acetamide (NAD or NAAm)	Synthetic auxin	Inexpensive, useful on late-season cultivars	Slower acting than NAA

cherry growers have used gibberellic acid to maintain vegetative growing points, especially as trees age. Recently, this strategy has been employed to thin peaches, where excessive flowering increases thinning costs and depresses fruit size. Table P3.2 lists the three thinning strategies for stone fruit, the growth regulators used, and their strengths and weaknesses. Some of the PGR registrations listed are restricted to particular geographic areas. Before using any of the materials listed in Table P3.2, a grower should consult the pesticide label for specific recommendations for the crop and farm location (American Crop Protection Association, 2001).

CONTROL OF PREHARVEST DROP OF FRUIT

Decreasing preharvest drop is probably the most cost-effective use of PGRs in apple production. As with chemical thinning, its use is widespread in the pome fruit industry. As fruit mature, the force required to remove them from the tree decreases. To prevent this, one of two methods is employed. The first, an older approach, is the applica-

TABLE P3.2. Potential time of thinning for stone fruit crops, materials used, and their strengths and weaknesses

Timing	Materials Used	Type of PGR	Strengths	Weaknesses
Summer	Gibberellin	Plant hormone	Greatest influence	Difficult to predict timing, cultivar specific
Bloom	Ammonium thiosulfate	Fertilizer	Inexpensive	Erratic performance
Postbloom	Ethephon	Releases ethylene	Later application	Gummosis, leaf abscission

tion of a synthetic auxin such as naphthalene acetic acid (NAA). This application is generally made during the harvest period and leads to a temporary suppression of fruit drop. This synthetic auxin directly affects the abscission zone, suppressing the enzymes needed to promote abscission. At the same time, auxin application leads to the production of "auxin-induced ethylene." Treated fruit remain attached but color and soften faster than untreated fruit. Consequently, storability of NAA-treated fruit is reduced. The other approach is to suppress the synthesis or action of ethylene in the attached fruit. Suppression of ethylene biosynthesis and preharvest drop began with the use of daminozide. This chemical was used successfully for about 20 years, until its registration was cancelled in the late 1980s.

About the same period daminozide registration was cancelled, scientists in S. F. Yang's laboratory at the University of California elucidated the ethylene-biosynthesis pathway (Adams and Yang, 1979). From that work, two PGR approaches emerged, one for suppressing ethylene biosynthesis and the other for suppressing its action in maturing fruit. The most widely used ethylene-biosynthesis inhibitor is aminoethoxyvinylglycine (AVG). With AVG usage, maturation is delayed and preharvest drop suppressed. As fruit can continue growing for an additional one to two weeks prior to the onset of endogenous ethylene production, AVG application can enhance fruit size.

Suppression of ethylene action occurs through treatment with 1-methylcyclopropene (MCP), which blocks ethylene's attachment to its binding site, thereby preventing ethylene action and ripening.

MCP usage is currently limited to postharvest applications to enhance storability, although it seems likely that an MCP formulation will be developed that can be used in the orchard.

The use of synthetic auxin as a chemical thinner, and again later in the season to prevent preharvest drop, presents an interesting conundrum. When a synthetic auxin is used to chemically thin, it is applied when fruit set is heavy. Since cell division is occurring, carbohydrate requirements are high in each fruitlet. Auxin application leads to a dramatic increase in fruitlet ethylene and epinasty (leaf curling and shoot bending below the tip) on the day of treatment (Walsh, Swartz, and Edgerton, 1979). In the next few days, decreased carbohydrate levels in the fruit are measurable along with a decrease in growth rate. These are followed by seed abortion and fruitlet drop, which typically occur about two weeks after application. In this situation, PGR usage builds on the naturally occurring waves of fruit abscission. When applied to control preharvest drop, growth-promoting hormones in the fruit are relatively low and size is much greater. Consequently, the major effect of an auxin application is to replace the growth-promoting signal needed to maintain an intact abscission zone. Although auxin-induced ethylene is produced, its role at that time is merely an effect on fruit quality. Figure P3.1 shows the differences that occur when auxin is applied for chemical thinning versus preharvest drop of apple.

SPECIALTY APPLICATIONS IN TARGETED SITUATIONS

Tree Development

Since endogenous auxin and cytokinin regulate lateral bud outgrowth in green shoots (apical dominance) and in woody shoots (apical control), they have been used successfully to improve tree architecture. In large top-dominant trees that require heavy pruning, water sprouts can be difficult to control following the removal of vigorous, upright limbs. If synthetic auxin is applied immediately following pruning, its application mimics apical dominance, and water sprout growth can be suppressed.

To bring young, high-density orchards into rapid production, tree training has taken precedence over pruning. In many cases, rapid growth does not provide the whorl of scaffold limbs where desired.

Spring Application of Synthetic Auxin   Fall Application of Synthetic Auxin

- Auxin-induced ethylene causes epinasty.
- Carbohydrate accumulation decreases.
- Fruit growth rate decreases.
- Seed abortion is stimulated.
- Fruitlet abscission occurs.

- Auxin-induced ethylene enhances ripening.
- Auxin maintains an intact abscission zone.
- Riper fruit remain attached to the tree.

FIGURE P3.1. A comparison of the differences in crop load, fruit size, and physiological effects that occur when synthetic auxins are applied in springtime as chemical thinners, or in fall to control preharvest drop of apple (*Source:* Illustration courtesy of Kathleen W. Hunt, University of Maryland, College Park, MD.)

To overcome apical control, cytokinin-plus-gibberellin mixtures are applied topically, to stimulate lateral bud outgrowth. In this case, the cytokinin leads to bud outgrowth and the gibberellin causes the shoot to develop vigorously (Williams and Billingsley, 1970). In laboratory situations, high concentrations of these chemicals are typically applied in a lanolin paste that allows uptake to occur slowly. In the field, applications of PGRs are made by hand, painting a solution of PGR directly onto the buds and tree bark. Commercial materials are typically mixed into wound-healing formulations used in arboriculture or are made by the grower through mixing with household latex paint (Harris, Clark, and Matheny, 1999).

Fruit Improvement

Another use of cytokinin-plus-gibberellin mixtures is to promote "typiness" in apple fruit. Typiness is the increase of the length-to-diameter ratio. It is thought that temperatures during the period of cell division affect typiness—cool temperatures favoring elongation and warm temperatures suppressing elongation. To mimic this effect, cytokinin-plus-gibberellin mixtures can be applied at the onset of bloom or during the bloom period.

Gibberellins are used by cherry and apple growers for various other fruit quality benefits. Since cherries are seeded, gibberellic acid treatment does not have a dramatic influence on size, although some effect occurs. The growth-promoting effects of gibberellin application lead to slightly firmer and brighter sweet cherry fruit. Apple producers use gibberellic acid to improve fruit finish, primarily in 'Golden Delicious'. This cultivar is prone to russet development that occurs in response to cool, wet, springtime weather, or as phytotoxity following pesticide applications. Early season applications of GA_3 reduce symptoms of russeting, leading to smoother fruit finish and improved marketability. Another minor use is to prevent cracking of the 'Stayman' cultivar. 'Stayman' cracking occurs well before harvest, rendering fruit unsalable. Multiple applications of GA_3 are made to improve skin elasticity, thereby reducing cracking later in fruit development.

The most widespread use of any PGR is the use of gibberellic acid in producing seedless table grapes. Grapes require seed fertilization to stimulate initial growth, but embryo abortion does not lead to abscission. In seedless cultivars, growers make multiple applications of gibberellin early in the growing season. The first application acts to elongate the rachis, and subsequent applications stimulate berry growth. To enhance these effects, vine girdling is used to reduce vegetative growth and partition photosynthate to the developing clusters.

Facilitation of Early or Mechanical Harvest

Ethylene-releasing compounds can be used in a variety of tree fruit crops during maturation and ripening. Ethephon application stimulates red color development, degradation of chlorophyll, flesh softening, and abscission. To enhance color development and advance harvest, ethephon application can be made close to the anticipated harvest date. Although this allows growers to pick early and capture higher prices, fruit have less shelf life and should not be stored for extended periods. If fruit are not picked within one to two weeks after application, fruit softening and drop can occur.

Tart cherries and processing apples that are mechanically harvested can also be treated with ethephon. Application to these crops is made primarily to allow mechanical harvesters to pick a greater proportion of the fruit in a single harvest. Ethephon has a slight effect on softening, but this is minor when compared to the ease of once-over mechanical harvest.

Major uses of PGRs to regulate cropping are chemical thinning, preharvest drop control, and the enhancement of fruit quality. PGRs also are used to reduce vegetative growth, enhance flowering, and affect fruit development. As understanding of plant growth and development improves, PGR usage becomes more precise and effective.

Related Topics: CARBOHYDRATE PARTITIONING AND PLANT GROWTH; FLOWER BUD FORMATION, POLLINATION, AND FRUIT SET; HIGH-DENSITY ORCHARDS; PLANT HORMONES; POSTHARVEST FRUIT PHYSIOLOGY; TRAINING AND PRUNING PRINCIPLES

SELECTED BIBLIOGRAPHY

Adams, D. O. and S. F. Yang (1979). Ethylene biosynthesis: Identification of 1-aminocyclopropane-1-carboxylic acid as an intermediate in the conversion of methionine to ethylene. *Proc. Natl. Acad. Sci.* 76:170-174.

American Crop Protection Association (2001). *Crop protection reference CPR 2001.* New York: C & P Press.

Dennis, F. G. (1973). Physiological control of fruit set and development with growth regulators. *Acta Horticulturae* 34:251-257.

Harris, R. W., J. R. Clark, and N. P. Matheny (1999). *Arboriculture,* Third edition. Upper Saddle River, NJ: Prentice-Hall.

Walsh, C. S., H. J. Swartz, and L. J. Edgerton (1979). Ethylene evolution in apple following post-bloom thinning sprays. *HortScience* 14:704-706.

Williams, M. W. and H. D. Billingsley (1970). Increasing the number and crotch angles of primary branches of apple trees with cytokinins and gibberellic acid. *J. Amer. Soc. Hort. Sci.* 95:649-651.

Zimmerman, R. H. (1972). Juvenility and flowering in woody plants. *HortScience* 7:447-455.

– 4 –

Plant Hormones

Christopher S. Walsh

Plant hormones can be defined as naturally occurring substances that regulate one or more developmental events in plants. To be classified as a plant hormone, a chemical is

- not one of the sixteen essential elements required for plant growth;
- synthesized from two or more of the following elements: carbon, hydrogen, oxygen, and nitrogen;
- not a sugar, vitamin, or enzyme cofactor;
- present and active in extremely low concentrations;
- endogenous, meaning that it occurs naturally within the plant; and
- synthesized in one or more locations, where it demonstrates activity, or is able to act at locations other than its site of synthesis.

Past research has identified five broad groups of compounds that meet these criteria. Modern hormonal theory recognizes that auxin, gibberellin, cytokinin, abscisic acid, and ethylene are endogenous hormones that work alone, and in concert, to regulate plant development. Auxin, gibberellin, and cytokinin are generally described as "growth-promoting hormones," while abscisic acid is known as an "inhibitor." Ethylene does not fit neatly into either category. As it is inextricably involved in plant senescence, some classify ethylene as an inhibitor. With recent advances in molecular biology, perhaps ethylene is better described as a "promoter of senescence." A synopsis of each hormone is presented below, including a brief history of its discovery, its sites of synthesis in the plant, its movement, and some of the developmental events regulated by the plant. The commercial applications of plant hormones and plant growth regulators are covered in the PLANT GROWTH REGULATION chapter.

AUXIN

Auxin, or indole-3-acetic acid, was the first plant hormone discovered. It was discovered by chance, in a human nutrition study conducted in the early twentieth century. One subject in that study had a disease that caused him to excrete large amounts of auxin in his urine. For a long period of time following that initial discovery, auxin was thought to be the sole plant hormone. Considerable research was conducted on the effects of exogenous auxin application on plant development and on internal levels of auxin like activity, with the goal of explaining the regulation of plant growth and development.

In plants, auxin is synthesized primarily in the apical meristem. It is actively transported basipetally, from the apical meristem toward the root. In vegetative tissues, this basipetal movement of auxin has three broad developmental roles that regulate plant architecture. The first is enhancing cell elongation and the differentiation of xylem elements below the apical meristem. The second is apical dominance, which is the suppression of lateral bud outgrowth at nodes below the apex in the current season's shoot. The third major function is the stimulation of root growth.

In perennials, auxin production affects development in woody tissues. Its production in the vascular cambium leads to the differentiation of xylem elements. It is also responsible for apical control, which is the regulation by the apex of shoot and spur development in the established woody structure of the plant. A secondary site of auxin synthesis is in developing leaves and fruit. Following synthesis, auxin moves basipetally from the leaf blade or fruit into the petiole or pedicel. There, its role is to maintain the abscission layer in a "young" state, so that the leaf or fruit remains attached to the plant. As leaf or fruit senescence occurs, auxin levels decline, and cells differentiate into an abscission zone, leading to leaf or fruit drop.

GIBBERELLIN

Gibberellin was first discovered in Japan, prior to World War II, by scientists who were studying rice infected with the "foolish-seedling" disease. The causal agent is a fungus that triggers plants to grow taller than normal and eventually lodge. The name of the family of gibberellin compounds was taken from the fungus. Nearly 100 endogenous compounds have been found in plants that contain gibberellin-like activity. To simplify gibberellin notation, they are listed as GA_n.

Over the years, we have recognized that most of the gibberellins identified in plants are precursors or degradation products that occur in the biosynthesis and metabolism of this hormone. Since different plant families possess different pathways of gibberellin synthesis and degradation, not all gibberellins are present in a given species. In fruit crops, the gibberellins of import are GA_3, GA_4, and GA_7. These appear to be present endogenously in plant and fruit tissues and can also be applied exogenously as plant growth regulators. For more information on gibberellin chemistry and endogenous changes that occur during fruit development, see Westwood (1993).

In fruit plants, gibberellins are synthesized primarily in young, expanding leaves just below the meristem, in developing seeds, and in rapidly growing fruit tissues. Unlike auxin, gibberellin can be translocated in any direction throughout the plant. Gibberellins play three developmental roles in fruit plants, corresponding to their sites of synthesis. In vegetative tissues, gibberellins play a major role in cell expansion that occurs just below the apical meristem. They stimulate cell expansion in the rapidly growing area below the meristem. As such, they are associated with internode length, and hence vegetative vigor. Since excessive vegetative growth is not desirable, gibberellin synthesis inhibitors such as paclobutrazol and uniconazole are of interest in fruit production. In fruit development, two events are regulated by gibberellin levels in the tissue: (1) the suppression of flower bud initiation in pome fruit by seed-produced gibberellins and (2) cell expansion during "final swell" of fruit with a double-sigmoid growth pattern, such as stone fruit.

Biennial bearing was a problem that vexed apple growers for centuries. Until the discovery of plant hormones, it was assumed that biennial bearing was required for nutritional reasons, so the tree could replenish reserves lost in cropping. M. A. Blake's classic experiments in hand thinning demonstrated that flower bud initiation is developmentally regulated in apple. If fruit are not removed early in the season, flowering is suppressed in the subsequent season. Chan and Cain (1967) demonstrated that the seeds are the source of the vegetative signal to the spur. Through exogenous applications, gibberellins were implicated as the hormone responsible for that vegetative signal.

In developing fleshy fruit with a final swell, gibberellins play a major role in determining final fruit size just prior to harvest. At that time, there is a dramatic increase in cell expansion and fresh weight gain in the fruit flesh. This correlates with increasing gibberellin responsiveness in the tissues.

CYTOKININ

Cytokinin was the third growth-promoting hormone discovered. Its discovery occurred when tissue-culture propagation of plants was in its infancy, in the early 1960s. At that time, scientists could maintain tissue in aseptic culture but were unable to stimulate cell division. Through trial and error, chemicals capable of stimulating cell division were eventually discovered in herring sperm and in coconut milk. These compounds were named cytokinins, a derivative of the word "cytokinesis," which means cell division.

Cytokinin is found in rapidly growing tissues, both vegetative and reproductive. It is produced in shoots and roots. Sachs and Thimann (1967) showed that cytokinins can counter the effects of apical dominance and induce lateral bud outgrowth. The research subsequently showed that these compounds can move isotope-labeled metabolites to their point of application, which was named "hormonally directed transport." Cytokinin can act as a growth factor and as an antisenescence hormone in leaves and fruits.

Three major uses of cytokinins occur in fruit production. The greatest is in micropropagation, where cytokinin levels are manipulated in culture to induce tissue proliferation. The second is in fruit tree development, where applications of cytokinin sprays or paints are used to stimulate lateral bud outgrowth and improve tree structure. In addition, cytokinins are used in apple development, as chemical thinners, and to alter the length-to-diameter ratio of apple fruit.

ABSCISIC ACID

Abscisic acid, or ABA, was the final plant hormone isolated and identified. It was originally studied as a growth inhibitor isolated from senescent tissue. It was given the name abscisic acid, as it was thought to cause organ abscission. Eventually, scientists realized that ethylene is the hormone that induces abscission, but that abscisic acid also has a number of roles in plant development. These roles are in plant water relations, dormancy, and in the later stages of seed development. Since there are no direct effects induced by ABA for enhancing tree development or productivity and it is expensive to apply in the field, there are no current uses for ABA in fruit production.

ETHYLENE

Anecdotal evidence of ethylene effects on plants was noted in the late nineteenth and early twentieth centuries. Escapes of gas from street lamps were observed to defoliate city trees. Subjective evidence, from kerosene heaters used to degreen citrus and cross ripening of commodities shipped in vessels with apples, suggested that a gaseous hormone exists. Burg and Burg (1965) found that ethylene is the "fruit-ripening hormone" when they used gas chromatography to demonstrate that an increase in ethylene evolution occurs prior to the respiratory climacteric in apples.

Although the primary focus of ethylene research has been in fruit maturation and ripening, it also plays a role in regulating vegetative growth. In dark-grown seedlings, ethylene controls plumular expansion and hypocotyl development. Ethylene is also produced in response to tissue wounding. Klein and Faust (1978) demonstrated an enhancement in ethylene levels in shoots subjected to severe summer tipping. Ethylene also increases in response to branch bending. As such, it suppresses extension growth and stimulates radial growth, stiffening limbs. When applied exogenously to apple trees, ethylene can suppress shoot growth and stimulate flowering in the subsequent season.

MODE OF ACTION

Our view of the mode of action of plant hormones has changed markedly in the past two decades. After the discovery of cytokinins, and the advent of widespread use of micropropagation, the hormonal balance theory was in favor (Dilley, 1969). That hypothesis pitted promoters against inhibitors in regulation of fruit and fruit tree growth and development. With recent advances in molecular biology, it is recognized that hormones are messengers that work to induce changes in response through specific developmental signals and pathways. Hormones facilitate the synthesis of new messages and enzymes to carry out developmental changes.

During the past century, scientists have identified five plant hormones. These highly powerful organic molecules regulate cell division, cell expansion, and cell differentiation. At the organ level, they also regulate numerous developmental processes. The results of basic research provide insights into the complex mechanisms controlling tree and fruit growth and development. From these studies, many commercial methods for improving productivity and quality have been developed.

Related Topics: CARBOHYDRATE PARTITIONING AND PLANT GROWTH; FLOWER BUD FORMATION, POLLINATION, AND FRUIT SET; HIGH-DENSITY ORCHARDS; PLANT GROWTH REGULATION; POSTHARVEST FRUIT PHYSIOLOGY; TRAINING AND PRUNING PRINCIPLES

SELECTED BIBLIOGRAPHY

Burg, S. P. and E. A. Burg (1965). Ethylene action and ripening of fruits. *Science* 148:1190-1196.

Chan, B. and J. C. Cain (1967). Effect of seed formation on subsequent flowering in apple. *Proc. Amer. Soc. Hort. Sci.* 91:63-68.

Dilley, D. R. (1969). Hormonal control of fruit ripening. *HortScience* 4:111-114.

Klein, J. D. and M. Faust (1978). Internal ethylene content in buds and woody tissues of apple trees. *HortScience* 13:164-166.

Sachs, T. and K. V. Thimann (1967). The role of auxins and cytokinins in the release of buds from dominance. *Amer. J. Bot.* 54:136-144.

Westwood, Melvin N. (1993). *Temperate-zone pomology: Physiology and culture.* Portland, OR: Timber Press.

– 5 –

Plant Nutrition

Dariusz Swietlik

All higher plants, with the exception of carnivorous ones, utilize nutrients of an exclusively inorganic (mineral) nature. Essential nutrients are defined as requisite for normal functioning of a plant's physiological and metabolic processes and cannot be substituted by other chemical elements or compounds. The essential nutrients are divided into macroelements, consisting of carbon (C), hydrogen (H), oxygen (O), nitrogen (N), phosphorus (P), sulphur (S), potassium (K), calcium (Ca), and magnesium (Mg), and into microelements, consisting of iron (Fe), manganese (Mn), copper (Cu), zinc (Zn), molybdenum (Mo), boron (B), and chlorine (Cl), with tissue concentrations that may be 100 to 10,000 times lower than those of macroelements. Nickel (Ni), cobalt (Co), sodium (Na), and silicon (Si) are difficult to classify at this time because the essential need for these elements in all higher plants has not yet been confirmed. Plants obtain most of their H, C, and O in the form of carbon dioxide (CO_2) and oxygen supplied by air and as water supplied by the soil. Many of all the other nutrients are absorbed from the soil, but plant foliage may also absorb small quantities of nutrients from rainwater, e.g., nitrate (NO_3), sulfur dioxide (SO_2), and others.

SOIL AS A RESERVOIR OF PLANT NUTRIENTS

Clay particles and some organic soil constituents carry a negative charge that allows them to adsorb positively charged nutrients (cations) such as NH_4^+, K^+, Ca^{2+}, Mg^{2+}, Fe^{3+}, Mn^{2+}, Cu^{2+}, and Zn^{2+}. Adsorbed cations can be exchanged, in chemically equivalent amounts, with those present in the soil solution. The adsorption of cations on soil particles minimizes their leaching losses to groundwater. The exchange replenishes the soil solution with nutrients depleted by root uptake. Soils have a low positive charge; hence, anions such as NO_3^-, BO_3^-, and Cl^- are easily leached, but $H_2PO_4^-$, MoO_4^{2-}, and SO_4^{2-} are specifically adsorbed or chemically react and precipitate and thus are effectively held by the soil.

Plant roots absorb nutrients primarily from the soil solution and secondarily from the soil exchange complex in contact with the root. Most of the nutrients reach a root's surface by mass flow or diffusion. The mass flow component is the product of the concentration of nutrients in the soil solution and the volume of plant water uptake. If a given nutrient is taken up faster than water, then that nutrient is gradually depleted in the immediate root vicinity, creating a diffusion gradient for its movement from bulk soil toward the root surface.

NUTRIENT ABSORPTION AND TRANSPORT

Root Uptake

The first step in ion uptake by roots involves crossing the cell wall of the epidermis (Figure P5.1). Further radial movement across the root cortex may proceed along two pathways. The first one involves diffusion of ions in the continuum of cell walls called the free space or apoplast. The cortex, however, is separated from the vascular cylinder (the stele) by a layer of cells called the endodermis whose anticlinal walls are impregnated with suberin (Figure P5.1). These suberized walls, known as the Casparian strip, are impervious to water and nutrients. To bypass this barrier, ions must cross the plasma membrane of the endodermal cells to enter the stele. The second pathway involves transport across the plasma membrane of epidermal cells and subsequent diffusion across the cortex and endodermis to the stele in the continuum of cell cytoplasm called the symplast. The cytoplasms of adjacent cells are connected via plasmodesmata, which are tubular extensions of the plasma membrane that traverse the cell wall (see Figure W1.1, WATER RELATIONS). Irrespective of the pathway followed, once in the stele, ions diffuse in the symplast toward xylem conducting elements (tracheids and vessels) (Figure P5.1). To enter the xylem, ions must exit the symplast and reenter the apoplast because xylem elements are dead cells. Ions are carried in

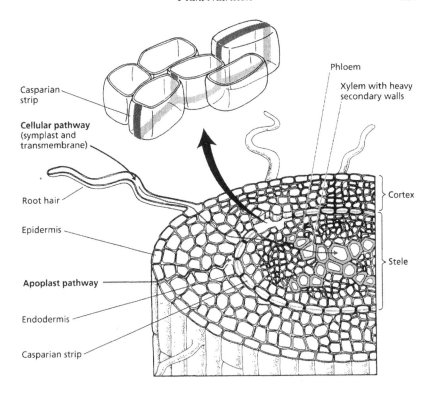

FIGURE P5.1. Pathways for nutrient and water uptake by roots (*Source:* Reproduced from Taiz and Zeiger, 1998, with permission from Sinauer Associates, Inc., Publishers.)

the xylem by the transpiration stream to the aboveground plant tissues (again, see Figure W1.1).

Transport Across Plasma Membrane

Ions are transported across membranes (plasmalemma or tonoplast) with the aid of transport protein systems called pumps, carriers, or channels. Transport down an electrochemical potential gradient is termed passive, whereas that proceeding against the electrochemical gradient is termed active. The three transport systems are (1) the primary active transport system, (2) the secondary active transport system, and (3) the passive transport system.

The primary active transport system includes H^+-ATPase (proton pump), which transports H^+ out of the cytoplasm into the apoplast, and Ca^{2+}-ATPase, which transports Ca^{2+} out of the cytoplasm. These processes require metabolic energy that is obtained by the hydrolysis of adenosine triphosphate (ATP) to adenosine diphosphate (ADP).

The secondary active transport system is driven by the electrochemical potential gradient for H^+ across the plasmalemma, called the proton motive force (PMF). The PMF consists of electric and concentration gradients generated by the proton pump in the process of extrusion of H^+ from the cell. The PMF-induced transport of H^+ across the membrane is coupled with an accompanying ion, which moves against its gradient of electrochemical potential. When the two ions move in the same direction, it is called symport, and the protein mediating that movement is termed a symporter carrier. When the two ions move in the opposite direction, it is called antiport or exchange, and the protein involved is termed an antiporter carrier. For example, K^+ is transported across the plasma membrane by a specific K^+-H^+ symporter when external K^+ concentrations are low, and specific antiporters mediate absorption of Cl^-, NO_3^-, and $H_2PO_4^-$.

The passive transport system involves ion movement across the plasmalemma or tonoplast via ion channels down an electrochemical gradient. Calcium is believed to enter the cell in this way. The electric gradient generated by the proton pump across the plasma membrane (-120 to -180 millivolts) is believed to be large enough to permit most microelements to enter passively into the cell via specific ion channels.

Long-Distance Transport

Mineral nutrients are transported upward in the xylem and downward or upward in the phloem. Prior to xylem loading and translocation to the aboveground parts, NO_3-N is reduced by nitrate reductase and incorporated into amino acids and amides. The process requires energy and C skeletons, which are generated by photosynthesis. All mineral nutrients are highly mobile in the xylem, but N, K, Mg, P, S, Cl, and Na also move easily in the phloem. Microelements are partially phloem mobile, whereas Ca is considered to be immobile.

Absorption by Leaves

Nutrients deposited on the leaf surface diffuse through the cuticle and cell wall of the leaf epidermis and may continue their inward migration in the apoplast or cross the plasmalemma and enter the cytoplasm. The leaf cuticle is covered with hydrophobic waxes, but cracks and discontinuities in these waxes open pathways for penetration of leaf-applied nutrients.

PHYSIOLOGICAL AND BIOCHEMICAL FUNCTIONS OF PLANT NUTRIENTS

Nitrogen is a building block for amino acids, amides, proteins, and alkaloids. Adequate N is essential for normal flowering, fruit set, and vegetative and fruit growth, but too much N induces excessive vegetative growth, poor color and quality of fruit, and reduced storage and shelf life.

Calcium plays an important role in binding polysaccharides and proteins that form the cell wall, stabilizing cell membranes, and regulating the activities of several enzymes in the cytoplasm, including those involved in the regulation of respiration rate. Low levels of Ca in fruit are associated with several physiological disorders of apple, pear, and other fruit, elevated susceptibility to postharvest diseases, and generally poor storage fruit quality.

Potassium is involved in protein synthesis, enzyme activation, stomatal movement, photosynthesis, and transport of photosynthates. The element acts as an osmoticum and maintains turgidity and growth of plant cells. Potassium applications to trees deficient in this element improve tree growth, fruit size, and apple red color, but an excess of K may exacerbate bitter pit in apples.

Phosphorus is an important structural element of deoxyribonucleic acid (DNA) and ribonucleic acid (RNA) and of phospholipid membranes. Also, it plays a role in hydrolysis of ATP to ADP and in the formation of ATP. Inorganic P regulates a number of enzymatic processes in the cell. Because fruit trees are very efficient in acquiring P from the soil, the deficiency of this element is usually not observed in orchards. Apple P correlates positively with fruit firmness and negatively with low-temperature breakdown in cold storage.

Magnesium occupies a central position in the chlorophyll molecule and also activates ribulose biphosphate carboxylase that plays a prominent role in photosynthesis. Magnesium forms a bridge between ATP and an enzyme, thus enabling phosphorylation and dephosphorylation, which are responsible for transfer of energy and activation of enzymatic processes. Through its effect on ribosomes, Mg plays a role in protein synthesis. Applications of Mg increase vegetative growth, fruit set, and fruit size in trees affected by severe deficiency of this element. Magnesium decelerates ripening and senescent breakdown of apples but also increases the incidence of bitter pit.

Iron is a constituent of cytochromes and nonheme iron proteins involved in photosynthesis, nitrite reduction, and respiration. Iron deficiency chlorosis occurs on fruit trees grown in neutral or high pH calcareous soils. Affected trees are less vigorous and unproductive.

Boron forms complexes with sugar derivatives and other constituents of cell walls. The element is involved in nucleic acid metabolism and in the process of cell division and elongation. Trees low in B suffer from poor fruit set because of the death of flowers, a condition known as "blossom blast." Deficient trees are less vigorous and develop small, deformed, and cracked fruit.

Manganese activates a number of important plant enzymes, some of which protect tissues from the deleterious effect of free oxygen radicals. Deficiency of Mn is usually associated with high pH soils and leads to leaf chlorosis and even tree defoliation. On acidic soils, where Mn availability is high, the element may be absorbed in excessive amounts and lead to the development of a physiological disorder on 'Delicious' apple trees known as internal bark necrosis or "measles."

Copper is a component of a number of important plant enzymes, particularly those involved in redox reactions. Copper deficiency may lead to severe shoot dieback, but such cases are rather rare.

Zinc is a component of a number of enzymes and acts as an enzyme cofactor. The element is required for the formation of tryptophan, which is a precursor for the auxin indole-3-acetic acid. Severe Zn deficiency drastically reduces shoot growth, narrowing and decreasing the size of terminal leaves and causing them to be bunched together at the shoot tips. The condition is known as "little leaf" or "rosette." Trees grown in high pH soils or in highly leached sandy soils are more likely to develop the deficiency.

MODERN TRENDS IN FRUIT TREE NUTRITION

Modern nutrient management practices rely on fine-tuning the application of nutrients to satisfy specific needs of different tree organs at times most beneficial from the standpoint of tree productivity and fruit quality. An improved understanding of how tree nutrient reserves are built up and mobilized leads to fertilizer practices that optimize yield and fruit quality while minimizing excessive vegetative growth. The use of different rootstocks with various abilities to acquire nutrients from the soil is being explored to solve tree nutritional problems via genetic means rather than fertilizer manipulations. A better understanding of the genetic control of plant nutrient uptake and translocation on a molecular level will open new frontiers for further improving the efficiency of mineral nutrient acquisition and utilization with the use of less fertilizer.

All these modern approaches to plant nutrition are aimed at minimizing or eliminating the environmental pollution that can potentially result from the use of fertilizers. Fertilizer practices will increasingly be assessed by their impacts on fruit nutritional value, concentrations of compounds beneficial to human health (nutraceuticals), and general health benefits.

Related Topics: ORCHARD FLOOR MANAGEMENT; ORCHARD PLANNING AND SITE PREPARATION; SOIL MANAGEMENT AND PLANT FERTILIZATION

SELECTED BIBLIOGRAPHY

Faust, Miklos (1989). *Physiology of temperate zone fruit trees.* New York: John Wiley and Sons.

Swietlik, Dariusz (1999). Zinc nutrition in horticultural crops. *Hort. Rev.* 23:109-178.

Swietlik, Dariusz and Miklos Faust (1984). Foliar nutrition of fruit crops. *Hort. Rev.* 6:287-356.

Taiz, L. and E. Zeiger (1998). *Plant physiology.* Sunderland, MA: Sinauer Associates, Inc.

– 6 –

Plant-Pest Relationships and the Orchard Ecosystem

Tracy C. Leskey

A pest is considered to be an organism in direct competition with humans for a valued resource. Pests have the potential to lower yields and reduce marketability of agricultural products. Five categories of pests can cause economically important damage in orchard ecosystems: arthropods, disease-causing pathogens, nematodes, vertebrates, and weeds. A particular type of damage can be caused by a single organism or by interactions among several organisms such as arthropod- or nematode-vectored diseases. Ultimately, damage in a particular orchard ecosystem is caused by organisms belonging to all categories, leading to economic losses.

ARTHROPODS

Arthropod pests can be divided into two categories, direct and indirect pests. Direct pests attack fruit and fruit buds, causing immediate injury. In some cases, damage is cosmetic, not affecting nutritional value or flavor but diminishing aesthetic quality for marketing purposes. However, injury by direct pests can reduce storability of fruit for processing and increase likelihood of secondary invaders. This type of damage is extremely important economically because fruit can be destroyed outright or rejected for fresh-market sale or for processing. Although damage by direct pests does not adversely affect tree vigor, indirect pests are extremely important to tree vigor, as they attack foliage, roots, limbs, or other woody tissues. Foliar-feeding arthropods reduce photosynthesis due to a reduction in either quality (e.g., mites) or quantity (e.g., leafminers) of photosynthetic leaf surface area. This, in turn, leads to a reduction in the amount of carbohydrates manufactured, ultimately resulting in future production problems, such as poor flower bud formation, reduced fruit set, and decreased fruit, shoot, or root growth. The feeding of insects on surfaces of woody tissue or within the cambial layer of roots, trunks, or limbs leads to declines in tree vigor and yield. Furthermore, this type of feeding facilitates secondary damage by opportunistic insects and diseases. Finally, some insects and arthropods may be important economically not because of any particular feeding damage they create but because of their ability to transport or transmit pathogenic or disease-causing organisms such as bacteria and viruses to deciduous fruit trees. For example, many species of blossom-visiting and flying insects transport the bacterium that causes fire blight, *Erwinia amylovora,* from infected to uninfected apple or pear trees. Such disease transmission may occur as the result of blossom visitation by insects carrying bacteria on their outer body surfaces (e.g., pollinating bees). Diseases, especially viruses, are readily transmitted by insects with specialized needlelike mouthparts (e.g., leafhoppers and aphids) used to probe plant tissue.

DISEASE-CAUSING PATHOGENS

Biotic diseases are the result of infection of a host plant by a pathogenic or disease-causing agent under environmental conditions favorable to the pathogen. The most common disease-causing pathogens important in orchard ecosystems include bacteria, fungi, and viruses. Disease severity is dependent on a number of factors, including tree cultivar, vigor and maturity, environment, and soil texture and quality. Generally, symptoms of pathogenic infection first appear in the region where the tree has been infected, such as bacterial and fungal infections associated with foliage, blossoms, fruit, and twigs. Such infections result in localized lesions on foliage and fruit. Trees are especially vulnerable to infection during bloom because blossoms provide an excellent entry way for infection, leading to wilt of blossoms and blossom stems. Foliar infections are important because photosynthesis declines due to destruction of leaf tissue; degeneration of chloroplasts due to infection also can lead to lowered levels of photosynthesis. Toxins associated with pathogenic infection can interfere with enzymes involved in photosynthesis. Furthermore, pathogenic infection of leaves by bacteria and fungi destroy foliar cuticle and epidermis, leading to increased transpiration rate and uncontrolled water loss, resulting in wilt, unless compensated by water absorption and translocation. The trunk and branches are a third area where pathogenic infections caused by fungi or bacteria commonly occur. Infection leads to canker formation that causes decline in food and water transport as cambial cells are destroyed, resulting in reduced growth, wilting foliage, and loss of crop and/or yield if fruit wood is destroyed. Infections in a fourth area, the root system and/or crown, by soilborne fungi or bacteria disrupt translocation of water and nutrients from soil. Here, the first obvious symptoms are not necessarily at the site of infection but instead are above ground. Some common symptoms include poor shoot growth, yellow foliage, dieback of terminals, and decreased productivity. Often these infections are confused with water-related stress. Bacterial or fungal infections of fruit itself can lead to economic losses, as fruit can be destroyed due to rot. Viral infections are generally insect vectored or spread from grafting and budding with infected rootstock or scion wood. The infection moves rapidly through the phloem, often leading to phloem degeneration and problems with transport of organic molecules produced in foliage following photosynthesis. Symptoms of viral infection include yellowing of foliage along leaf veins, premature leaf drop, and reductions in growth, yield, and fruit quality. Ultimately, physiological responses induced by pathogenic organisms infecting deciduous tree fruit can affect size, shape, appearance, and overall quality of fruit; induce distortions of leaves and premature leaf drop; cause dieback; reduce tree vigor; and lead to death of trees.

NEMATODES

Nematodes are unsegmented roundworms belonging to the phylum Nematoda. Those considered to be pests of plants are active mainly in moist habitats and feed on roots. Some species are ectoparasites, feeding on root surfaces, while others are endoparasites, feeding internally on root tissues. Injection of saliva by nematodes induces distortion of roots, including galling, stubby roots, lesions, and stunting, leading to decreased translocation of water and nutrients. Feeding by nematodes also predisposes trees to secondary infection by pathogens, and some species transmit viruses. The most common nematode pest in tree fruit is the root lesion nematode, *Pratylenchus penetrans;* this species damages roots by feeding and intracellular migration, leading to cortex damage and promotion of rot. Symptoms of nematode attack in tree fruit include stunting, chlorosis, wilting and curling of leaves and stems, heavy flowering leading to a large crop of small fruit, delayed or uneven maturation of fruit crops, and fruit drop.

VERTEBRATES

Vertebrate pests belong to the phylum Chordata and are easily identified if seen, but in general damage will be present and the animal will not. The most common vertebrate pests are birds, rodents (such as voles and rabbits), and deer. Birds generally attack ripe fruit; damage from pecking lowers marketability and leads to secondary entry of pathogens and insects. However, some birds feed on fruit buds as well. Damage by rodents such as voles leads to girdling of seedlings and young trees and damage to roots, while damage by rabbits includes feeding on buds and gnawing on bark as well as clipping off small branches. Most damage occurs in winter when other food sources are scarce. Browsing by deer also can be most damaging in the winter. The most susceptible trees include dwarf, semidwarf, and young standard trees. Trees may be stunted and fruit production may be affected by browsing on terminal and fruit buds.

WEEDS

Weeds are plants that compete in orchard ecosystems for soil moisture, nutrients, and sunlight. Weeds characteristically have rapid seed germination and seedling growth as well as root systems that have deeply penetrating and abundant fibers. Often weeds are better adapted than the tree fruit crop to a particular region. Weeds are of greatest concern in young orchards where they can severely reduce rapid growth of young fruit trees, leading to stunted trees and delays in flowering and cropping.

It is extremely important to understand plant-pest relationships within a class of pests as well as the interactions that can occur between or among classes when designing integrated control programs for orchard ecosystems. Treatment strategies aimed at controlling one pest can exacerbate problems with another. Conversely, careful orchard design and choice of treatment strategies can minimize problems associated with pests of several classes.

Related Topics: DISEASES; INSECTS AND MITES; NEMATODES; ORCHARD FLOOR MANAGEMENT; SUSTAINABLE ORCHARDING; WILDLIFE

SELECTED BIBLIOGRAPHY

Hogmire, H. W. Jr., ed. (1995). *Mid-Atlantic orchard monitoring guide,* Pub. NRAES-75. Ithaca, NY: Northeast Regional Agric. Engin. Serv.

Howitt, A. H. (1993). *Common tree fruit pests,* NCR 63. East Lansing, MI: Michigan State Univ. Exten. Serv.

Ogawa, J. M. and H. English (1991). *Diseases of temperate zone tree fruit and nut crops,* Pub. 3345. Oakland, CA: Univ. of California, Div. of Agric. and Nat. Resources.

Ohlendorf, B. L. P. (1999). *Integrated pest management for apples and pears,* Second edition, Pub. 3340. Oakland, CA: Univ. of California, Div. of Agric. and Nat. Resources.

Travis, J. W., coordinator (2000). *Pennsylvania tree fruit production guide.* State College, PA: Pennsylvania State Univ., College of Agric.

Postharvest Fruit Physiology

Christopher B. Watkins

Postharvest fruit physiology describes the interaction of the physiological and biochemical events associated with ripening and senescence. Fruit continue to function metabolically after harvest, but in the absence of carbohydrates, nutrients, and water supplied by the tree. In some cases, the fruit are eaten immediately. However, handling, storage, and transport technologies are usually used to maintain the rate of fruit ripening and therefore quality. The postharvest period hence involves the appropriate management of stress to minimize metabolic rates and/or enhancement of injurious metabolic processes.

The botanical definition of a fruit is "a seed receptacle developed from an ovary," but a range of fruit types exist (Kays, 1997). The fleshy part of the apple and pear develops from the accessory tissue of the floral structure, while the drupe fruit, characterized by the peach and apricot, develops from the mesocarp. While it is not surprising that differences in metabolism occur among fruit types, the central metabolic pathways of glycolysis, the tricarboxylic acid cycle, and the mitochondrial electron transport chain are common to all fruit. Fruit ripening is characterized by many events, the most important of which for perception of fruit quality in the marketplace, are changes of texture, color, and flavor. Other ripening-associated events include seed maturation, fruit abscission, changes in respiration rate, ethylene production, alterations in tissue permeability and protein contents, and development of surface waxes. Ripening changes involve a series of coordinated but loosely connected biochemical pathways. Many of these require anabolic processes that need energy and carbon skeleton building blocks supplied by respiration.

This chapter is restricted to the consideration of physiology and biochemistry of fruit ripening as it pertains to respiration and ethylene production, texture, color, and flavor because of the importance of these factors in affecting consumer acceptability of temperate tree fruit. Excellent overviews of postharvest physiology are available in books by Kays (1997), Knee (2002), and Seymour, Taylor, and Tucker (1993).

RESPIRATION AND ETHYLENE PRODUCTION

Fruit are classified as climacteric or nonclimacteric, based on the presence or absence of a respiratory increase during ripening. The climacteric rise is associated with increases in internal concentrations of carbon dioxide and ethylene, and of respiration and autocatalytic ethylene production. For example, in apples, respiration and ethylene production may increase by 50 to 100 percent and 1,000-fold, respectively. Climacteric fruit also can be differentiated from nonclimacteric fruit by their responses of respiration and/or ethylene production to exogenous ethylene or its analogues, such as propylene. In a climacteric fruit, ethylene advances the timing of the climacteric, autocatalytic production continues after removal of ethylene, and in contrast to a nonclimacteric fruit the magnitude of the respiratory rise is independent of the concentration of applied ethylene. Thus, timing of the respiratory increase and ripening of climacteric fruit is advanced by exposure to ethylene. The initiation of ripening and subsequent development of positive quality factors and negative storability factors are often associated with the climacteric. Most temperate tree fruit are climacteric, the major exception being the cherry. In the cherry, the respiration rate declines during growth and development and remains at low levels during ripening. Nonclimacteric fruit, in contrast to climacteric fruit, do not ripen after harvest.

As with any classification system, these categories are an oversimplification. Some fruit, such as nashi pears, have cultivars that are climacteric or nonclimacteric. Nonripening mutants of nectarines and other fruit have been identified, as have "suppressed climacteric" plums that do not produce sufficient ethylene to coordinate ripening but show characteristic responses to propylene. Also, differences in physiology can occur within climacteric fruit—early season apple cultivars, for example, tend to have much higher rates of ethylene production and respiration and ripen faster than late-season cultivars.

Fruit-ripening classifications are not measures of perishability, as evidenced by cherries that deteriorate quickly after harvest compared with apples that can have long storage lives. However, harvest of fruit before the climacteric and application of postharvest handling techniques, such as low-temperature and controlled atmosphere storage, reduce or eliminate the respiratory climacteric and generally reduce respiration rates. Low storage temperatures are the primary means of reducing metabolic rates, but the safe temperature range is influenced by susceptibility of the fruit to chilling injury. Most apple and pear cultivars are resistant to development of chilling injury, while stone fruit are more sensitive to injury.

The effects of temperature on ethylene metabolism are different from those on respiration. Low temperatures delay ethylene production in some apple cultivars but enhance it in others. Some pears require a chilling period to induce ethylene production and proper ripening.

TEXTURE

Changes in texture, perceived as crispness, juiciness, and hardness, occur during fruit ripening, but the type and extent of change varies greatly by species and cultivar. Temperate fruit can be divided into two groups—those which soften considerably to melting texture, such as the pear, plum, and peach, and those which soften only moderately and retain a crisp fracturable texture, such as the apple and nashi pear. Within both groups, variations occur. These include the peach, in which freestone types soften to a greater extent than clingstone types. Also, apple cultivars such as 'Cox's Orange Pippin' and 'McIntosh' lose texture rapidly after harvest in contrast to 'Honeycrisp' and 'Fuji' that maintain texture for extended periods.

Fruit flesh consists mainly of thin-walled parenchyma cells with considerable intercellular space between them. The primary cell wall consists of cellulose microfibrils embedded in a matrix of other polysaccharides and proteins. A pectin-rich middle lamella region connects the adjacent cell walls. Wall-to-wall adhesion, which is a function of the strength of the middle lamella, the area of cell-to-cell contact, and the extent of plasmodesmatal connections, is considered to be the major factor affecting fruit texture.

Loss of texture can be caused by loss of turgor and degradation of starch, but the primary cause is disassembly of the primary wall and middle lamella. However, the mechanisms of this disassembly during ripening are still unclear, despite intensive research at the chemical, microscopic, enzymatic, and molecular levels. Much less research attention has been given to temperate fruit compared with the tomato, often used as a model system. Nevertheless, those fruit which soften to melting textures are characterized by pronounced cell wall swelling and pectin solubilization; cell-to-cell adhesion is poor and minimal cell disruption occurs. In contrast, no cell wall swelling and little pectin solubilization occurs in fruit that remain crisp; cell-to-cell adhesion is strong and cell walls rupture.

Cell wall hydrolases that have been studied include pectin methylesterase (PME), endo-β-1,4-gluconase (cellulase), β-galactosidase, and xyloglucan endotransferase (XET), but especially polygalacturonase (PG) because of the pronounced pectin solubilization that occurs during ripening. Moreover, initiation of fruit ripening is often associated with expression of genes encoding PG and its increased activity. Although recent transgenic studies with tomato have not supported a straightforward relationship between PG activity and softening, close correlations between the factors have been repeatedly documented. Examples include the apple and cherry where endoPG activity is low or undetectable, and little pectin depolymerization occurs. Freestone peaches have high activities of both endo- and exoPG and pronounced pectin solubilization, whereas clingstone peaches that soften to a lesser extent than freestone peaches have less pectin solubilization and lower endoPG activity. Increased pectin solubilization and PG activity occurs in winter pears only after the chilling requirement during storage has been met.

Failure to soften properly has been related to impaired cell wall metabolism. Development of mealiness in apples and pears, and woolliness in peaches, is associated with separation of parenchyma cells, an absence of cell fracturing, and an absence of free juice on the cell surfaces. It has been proposed that development of woolliness results from enhancement of PME activity and inhibition of endoPG activity in fruit kept at low storage temperatures.

COLOR

The major pigment changes that occur during ripening of fruit are losses of chlorophyll and either the unmasking of previously synthesized, or the synthesis of, pigments such as carotenoids and anthocyanins. Although chlorophyll loss and pigment synthesis are coordinated during ripening, the two events are not directly related or interdependent. In temperate tree fruit, the predominant red color pigments are anthocyanins. The timing, rate, and extent of color change vary greatly by fruit type and cultivar and can be affected by both preharvest and postharvest factors.

Chlorophylls are sequestered in the chloroplasts as two predominant forms, chlorophyll a and chlorophyll b. Chlorophylls are insoluble in water. Although the mechanism of chlorophyll degradation is not fully understood and may involve both enzymatic and chemical reactions, chlorophyllase activity is thought to play a major role. Most temperate tree fruit lose chlorophyll during ripening, although some apple and pear cultivars remain green. It can be difficult to separate the initial decline of chlorophyll, on a surface area basis, resulting from fruit expansion from the subsequent yellowing that results from chlorophyll breakdown. Chlorophyll loss can be accelerated by increased ethylene production by the fruit, and yellowing is typically associated with softening and reduced market appeal. However, change of the background color is used as a harvest guide for bicolored cultivars such as 'Gala', 'Braeburn', and 'Fuji'. Exposure to ethylene after harvest can result in rapid loss of chlorophylls.

Carotenoids are a large group of water-insoluble pigments associated with chlorophyll in the chloroplast. Carotenoids are terpene compounds derived from acetyl-CoA via the mevalonic acid pathway. During ripening, carotenoid production increases, producing yellow to red pigments, as the chloroplasts are transformed to chromoplasts. The carotenoids found in developing apple fruit include β-carotene, lutein, violaxanthin, neoxanthin, and cryptoxanthin. Concentrations of lutein and violaxanthin increase, while levels of β-carotene decrease, during ripening. Relatively less is known about ripening-associated changes of carotenoids in other temperate tree fruit. Studies in peaches show that only traces of carotenoids are present in ripe fruit.

Anthocyanins are flavonoid compounds and are synthesized from phenylalanine. Anthocyanins are water soluble and accumulate in the vacuoles, producing pink, red, purple, and blue colors of fruit. A typical anthocyanin is cyanidin-3-galactoside, which is largely responsible for the color of apples and pears. Anthocyanins are distributed throughout the fruit, as in sweet cherry cultivars, or restricted to epidermal and subepidermal tissues, as in apples, pears, plums, and nectarines. Anthocyanin concentrations are affected by cultivar, and because of marketing pressures, extensive selection of early red color strains of apple cultivars has occurred. Preharvest factors that influence anthocyanin production include light quality and quantity and temperature. In apples, preharvest sprays of ethylene-producing compounds are sometimes used commercially to stimulate color development. Color changes during ripening of apple depend mainly on simultaneous disappearance of chlorophyll a and b. Anthocyanin concentrations increase little after harvest, even in the presence of ethylene, with apparent changes in appearance, such as redder fruit, being due to degradation of chlorophyll.

FLAVOR

Flavor is a function of two primary attributes, taste and odor. Whereas taste is related to perception of sweetness, sourness, bitterness, and saltiness by the taste buds in the mouth, odor depends on the contributions of specific odor volatiles perceived by the olfactory receptors in the nose. The flavor, and therefore consumer acceptance, of fruit is a complex interaction between the concentrations of sugars, organic acids, phenolics in some fruit, and volatile compounds.

Sugar concentrations increase during fruit ripening and are major determinants of sweetness. In nonclimacteric cherries, which cannot ripen after harvest, sugar concentration is solely a function of translocation of carbohydrates into the fruit before harvest. Therefore, fruit quality is greatly affected if harvest occurs before adequate sugar accumulation has occurred. In climacteric fruit, sugar increases also occur because of carbohydrate translocation, but hydrolysis of stored carbohydrates, especially starch in apples and pears, can be a major contributor to increased sugar concentrations. These fruit can reach acceptable sugar levels and flavor during ripening off the tree.

Acidity levels are important factors in the flavor of many temperate tree fruit. The major organic acids vary by fruit type, for example, malic acid being present in apples and cherries, malic and quinic acids in pears, and malic and citric acids in peaches and nectarines. The concentrations of these acids typically decline during ripening and are utilized as respiratory substrates and as carbon skeletons for synthesis of new compounds. Organic acid concentrations are greatly in excess of those required for energy during ripening, but they may decline markedly during ripening.

Astringency in fruit is determined by concentrations of phenolic compounds. These are usually derived from phenylalanine via cinnamic and coumaric acids. Astringency can be a characteristic of certain apple cultivars. In peaches, research indicates that low-quality fruit have higher concentrations of phenolics such as chlorogenic acid and catechin than high-quality fruit, but no changes are detected during ripening. The overall taste of a fruit can be affected by the balances among sugars, acids, and phenolics, rather than the concentrations of each one alone.

The aroma of each fruit results from distinct quantitative and qualitative differences in compositions of volatile compounds produced during ripening. The major classes of flavor compounds are aldehydes, esters, ketones, terpenoids, and sulfur-containing forms. Their respective biosynthetic pathways are diverse, including those involved in fatty acid, amino acid, phenolic, and terpenoid metabolism. Increases in volatile production are often, but not always, associated with ethylene production. While large numbers of individual volatiles have been identified in fruit, relatively few make up the characteristic aroma perceived by the consumer. In apple fruit, for example, over 200 volatiles have been identified, but ethyl 2-methylbutyrate is responsible for much of the characteristic apple odor. Important "character" volatiles may occur in very low concentrations. In some cases, a single or few volatiles that make the "character" aroma have been identified, while in others, aroma is made up of a complex mixture of compounds that cannot be reproduced.

The ripening of fruit is a complex phenomenon that varies greatly among and within fruit types. Understanding the physiology of these fruit is critical to application of handling and storage protocols that will result in acceptable quality in the marketplace. Knowledge of respiratory and ethylene responses, for example, allows harvest decisions and application of correct storage temperatures that will minimize unwanted ripening changes. Although the successes of horticultural industries around the world are evidence of tremendous progress, serious issues exist with fruit quality in the marketplace. Unfortunately, appearance of the fruit is one of the last characteristics to be lost; a fruit may appear attractive but be soft and flavorless. Moreover, some of the methods used to prolong storage life, especially low-oxygen storage, appear to affect detrimentally recovery of flavor volatiles.

Related Topics: FRUIT COLOR DEVELOPMENT; FRUIT MATURITY; HARVEST; PACKING; PHYSIOLOGICAL DISORDERS; PROCESSING; STORING AND HANDLING FRUIT

SELECTED BIBLIOGRAPHY

Kays, S. J. (1997). *Postharvest physiology of perishable plant products.* Athens, GA: Exon Press.

Knee, M., ed. (2002). *Fruit quality and its biological basis.* Sheffield, UK: Sheffield Academic Press.

Seymour, G. B., J. E. Taylor, and G. A. Tucker, eds. (1993). *Biochemistry of fruit ripening.* London, UK: Chapman and Hall.

– 8 –

Processing

Mervyn C. D'Souza

Processed fruit products, especially sauce, slices, and juice, play an important role in the utilization and marketing of temperate tree fruit. Approximately 50 percent of the pome and stone fruit produced globally are processed.

APPLE CULTIVARS AND QUALITY CHARACTERISTICS

Major apple cultivars used for processing include 'York Imperial', 'Golden Delicious', 'Rome', 'Delicious', 'Granny Smith', 'McIntosh', 'Idared', 'Greening', 'Stayman', 'Jonathan', 'Empire', and 'Cortland'. In the last few years, newer cultivars such as 'Gala', 'Fuji', 'Braeburn', 'Jonagold', and 'Crispin' have also been used for processing. Each apple cultivar has unique storage and processing characteristics. For instance, 'York' apples store well and have firm texture, yellow flesh, and high acidity. 'Romes', in comparison, have lighter flesh color and softer texture. Some of the more important characteristics of a good processing apple cultivar are long storage potential, good size, round shape, good firmness and texture, sweetness, pleasant flavor, and high acidity. Growers and processors work together to determine the suitability of a particular cultivar for processing. Most of the newer cultivars now grown are dual-purpose apples that can either be sold fresh or used for processing.

APPLE SORTING

Apple processing is now a year-round operation. Therefore, fruit maturity and condition at the time of receiving is important in determining the disposition of each load of apples. Maturity is important from a processing and storage standpoint. Criteria used to determine maturity include firmness, soluble solids, starch content, and skin and seed color. Fruit received in an immature condition result in poor processed quality and develop disorders in storage. Fruit received in an overripe condition have short storage life. As apples are delivered to the processor over the fall harvest periods, they are sorted based on their storage potential and suitability for applesauce, slices, or juice. Apples that are of optimum maturity are placed in controlled atmosphere storage for use in the spring and summer months. Others are placed in regular storage or used directly for processing. Apples sorted for slice production are of the highest quality with respect to texture, shape, size, external and internal defects, lack of bruises, and insect or disease blemishes. Apples sorted for sauce or juice are of a lesser quality, since these products can tolerate fruit with minor blemishes.

In the United States, payment to growers for processing apples is determined by cultivar, quality, and an inspection carried out by an official of the U.S. Department of Agriculture (USDA). Additional information on grading of processing apples can be found in grading and inspection manuals developed by government agencies such as the USDA.

APPLESAUCE PRODUCTION

Firm, mature apples produce the best-quality applesauce. Usually two to five cultivars of apples are blended for uniformly high-quality sauce. Apples are transferred into a hopper where the cultivars are blended. Next they are run over sizing chains to grade out the smaller apples, which are used for juice. Apples are then inspected for decay and other blemishes such as hail marks, deep bruises, and scab lesions. Fruit are washed with potable water and peeled and cored using high-speed peeling machines. Peeled apples are further inspected for dark bruises, stems, or calyx remnants. Peeled apples may be conveyed by flume with an antioxidant such as citric acid or ascorbic acid to prevent the fruit from browning. Peeled apples are either chopped or diced and then processed in a cooker with live culinary steam at temperatures between 100 and 114°C, depending on the apple texture and the sauce characteristics desired. Other ingredients such as cinnamon, flavoring, purees, and sweeteners may also be added to the cooker, depending on the type of sauce being produced. Cooked apple material is run through a finisher screen with pore size ranging from 0.056 to 0.238 centimeters to remove seeds, carpel, peel, and other defects. Finished sauce is collected in a kettle and stirred constantly. Sauce is then filled into containers at a minimum temperature of 87.5°C to control microorganisms that cause spoilage. These containers are allowed a two- to three-minute sterilization time to sanitize the caps and headspace prior to cooling. Cooled sauce should be in the 37.4 to 42.9°C range for color and quality retention. Containers are then labeled, placed in cartons, and stacked on pallets. Pallets are moved to finished goods warehouses for storage and subsequent shipment.

APPLE SLICE PRODUCTION

Firm apple texture is a prerequisite for high-quality processed slices. Unlike applesauce, only one cultivar is used at a time for packing apple slices. 'York Imperial', 'Rome', 'Golden Delicious', 'Fuji', and 'Granny Smith' are good cultivars for slice production. All operations up to peeling are identical to applesauce production as discussed earlier. At the peelers, in addition to peeling and coring, apples are sliced using eight to 12 cuts, depending on fruit size and slice size desired. Slices are then flumed in an antioxidant solution to prevent browning and run over a shaker screen to remove fines and other small pieces. Slices with defects, including deep bruises, scab, or attached peels are inspected out. Slices are next retorted and blanched to remove the air and are partially cooked with steam. Blanched slices are placed in a container with syrup or water and sealed using a closing machine or a seamer. Closing temperature should be at least 76.5°C. Cans are cooked to maintain a center can temperature of 82°C. Cans are then cooled to a maximum of 56.7°C and subsequently labeled and cased as described under applesauce.

APPLE JUICE PRODUCTION

Apple cultivar selection is important in the manufacture of apple juice. Two to three apple cultivars are commonly chosen for blending. Cultivars selected should possess certain necessary sugar, acid, flavor, and texture characteristics. During production, apples are transferred into a hopper where they are mixed. They are then inspected for decay and other defects and washed with potable water. Apples are next chopped to a pulp, which is heated to 12.7 to 15.4°C for enzyme treatment. Mash is treated with enzymes for enhanced juice yields. Treated mash is pressed and run over a screen to remove fine apple particles. Juice is heated to temperatures in excess of 92.4°C and cooled to 56.7°C for further enzyme treatment. Enzymes are added to depectinize the juice and to remove starch. Pectin- and starch-free juice is then filtered and chilled to –1.1 to 1.7°C. Prior to bottling, juice is heated and filled in containers at 83.1 to 84.2°C. The containers are then cooled to 31.9 to 37.4°C prior to labeling, casing, and palletizing.

QUALITY ASSESSMENTS

Government standards are commonly used to perform quality assessments of processed apple products. USDA quality criteria for "A" grade are briefly as follows:

Applesauce

Color of regular applesauce should be bright, uniform, and typical of the cultivar(s) used with no discoloration due to oxidation or scorching. Consistency or flow of sauce should not exceed 6.5 centimeters, and free liquid should not be more than 0.7 centimeters, as measured using standard USDA flow charts. Sauce should be relatively free from defects, including dark stamens (not more than three), seed particles, discolored apple particles, carpel tissue (not more than 0.5 square centimeters), and medium- and dark-colored particles (not to exceed 0.25 square centimeters). Finish or graininess should be evenly divided and not be lumpy, pasty, or salvy. Sauce flavor should be tart to sweet and free from astringency. In grading applesauce, each of the previous criteria is given a score ranging from 18 to 20. For the sauce to be graded "A," the total score received should be at least 90. Scores between 80 and 90 would make the sauce of "B"-grade quality.

Apple Slices

Color is one of the important quality parameters for slices. For "A" classification, apples with good color that is uniformly bright both internally and externally and characteristic of the cultivar are given a score of 17 to 20 points. Size should be uniform to obtain a score between 17 and 20 points. This requires that at least 90 percent of the drained weight of the product consists of whole or practically whole slices. Canned slices that are practically free from defects may be given 17 to 20 points. This means that any extraneous matter present does not materially affect the appearance or eating quality of the slices. Finally, slices that possess a good character are assigned a score of 34 to 40 points. Good character implies that the slices possess a reasonably tender texture and have less than 5 percent mushy apples.

Apple Juice

For apple juice to be graded as "A," color of the product has to be bright and sparkling. Points between 18 and 20 are then assigned. Juice has to be free from defects, including amorphous sediment, residue specks, pulp, and other particles. Scores assigned range from 18 to 20. Flavor has to be good to qualify for a score of 54 to 60 for "A" grade. Samples receiving a total score of at least 90 points qualify for "A"-grade apple juice.

In addition to the quality parameters described in this chapter, other criteria such as brix, acidity, drain weight, and pH may be used to grade the various processed products. Detailed descriptions of quality assessments can be found in the USDA standards guides for each product.

OTHER TEMPERATE FRUIT

Peaches are processed into pie fill or other products such as slices and dices. Fruit are transferred into a hopper and then inspected for decay and superficial defects. Peaches are next run over a sizer to grade out extremely large or small fruit. They are cut in half and pitted by machine. Pitted halves are lye peeled using caustic potash (potassium hydroxide) and sliced. Slices are combined with pie fill slurry or sugar syrup and processed in a cooker. Cans are then cooled, labeled, and cased as described earlier.

Tart cherries are processed in a slightly different fashion. After harvest and prior to processing, cherries are cooled in tanks filled with ice. Cooling cherries helps pit removal and improves yields. Cherries to be processed are inspected for stems, leaves, and other extraneous material. They are then pitted by machine. Pitted cherries are further inspected for loose pits either manually or electronically. Fruit are filled in cans combined with slurry or water depending on the product. Cans are sealed and processed in a cooker. Cooled cans are then labeled, cased, and palletized.

Pears, plums, and apricots are some of the other temperate fruit that are processed. In addition to the products described earlier, these fruit may also be processed into juice, concentrate, and purees or as frozen slices or dices based on availability and customer requirements.

Processing plays a key function in the usage of temperate fruit crops. Products such as sauce, slices, juice, pie fills, and purees are commonly produced. Current processing technology enables efficient utilization of the fruit and provides the customer with a choice of healthy, high-quality products.

Related Topics: FRUIT MATURITY; HARVEST; PACKING

SELECTED BIBLIOGRAPHY

Downing, Donald L. (1989). *Processed apple products.* New York: Van Nostrand.
Processing Apple Growers Marketing Committee (2000). *Apple crop statistics and marketing analysis.* Michigan: Michigan Agric. Coop. Marketing Assoc., Inc.
Rowles, Kristin L., Brian Henehan, and Gerald White (2001). *New potential apple products: Think afresh about processing, An exploration of new market opportunities for apple products.* Ithaca, NY: Cornell Univ.
U.S. Apple Association (2000). *Apple crop outlook and marketing conference proceedings.* McLean, VA: USAA.

Propagation

Suman Singha

Vegetative, or clonal, propagation is the very basis of the tree fruit industry. For instance, 'Golden Delicious' apple trees found the world over have as their parent the single tree found on the Mullins farm in Clay County, West Virginia. These trees have been propagated by budding or grafting onto a desired rootstock and have the same genetic makeup (excluding bud sports or mutations) as the parent tree. This is true of every cultivar of temperate tree fruit. Other methods of propagation, including cuttings, layering, and tissue culture, may also be used to produce clonal rootstocks or self-rooted trees. Sexual propagation through seeds is used primarily for the production of rootstocks.

SEXUAL PROPAGATION

Sexual propagation is used almost entirely for the production of rootstocks that will be grafted or budded. It cannot be used to propagate the parent tree, as the progeny will not be true to type and will be unlike the parent.

For sexual propagation, seeds extracted from mature fruit are cleaned to remove any adhering fruit pulp prior to being stratified. Seeds of temperate tree fruit have an endodormancy and do not germinate if they are directly planted at ambient temperature; the seeds must be allowed to undergo a period of stratification to overcome this dormancy. Seed dormancy is a survival mechanism that ensures that in nature the seeds will not germinate when the fruit fall to the ground in late summer but instead will germinate after adverse winter conditions are past. Seeds can be stratified naturally by planting them outdoors in nursery beds; their dormancy requirements will be satisfied in the moist soil and the cold winter temperatures. If this approach is used, it is important to test the viability of the seed lot that is being planted. This can be done using embryo excision or using 2,3,5-triphenyltetrazolium chloride. These tests allow seed viability to be ascertained in a very short time period. Knowing the seed viability is important if a good stand of seedlings is to be expected in the following spring.

Seeds can be artificially stratified by placing them in moist media and holding them at 4°C. It takes approximately 60 to 90 days to fulfill the dormancy requirement, and the seeds can then be planted at the desired location.

VEGETATIVE PROPAGATION

Vegetative propagation is done through a number of methods, including cuttings, layering, grafting and budding, or tissue culture. The ease of propagating plants by each of these methods varies depending on species and cultivar.

Cuttings

Stem, leaf, or root cuttings are most commonly used to propagate herbaceous plants; however, stem cuttings have been used to propagate clonal rootstocks of *Prunus* species and also self-rooted trees, especially for peaches. Depending upon the species, hardwood cuttings collected during the dormant period, softwood cuttings in spring, or semihardwood cuttings in fall may be used. For semihardwood cuttings, for example, the process requires collecting the current season's growth in late summer, cutting the shoots into sections approximately 15 centimeters long, and removing all but the four or five leaves at the upper portion of the cutting. The bottom of the cutting should be lightly wounded, treated with a rooting hormone, and then planted in moistened growing media. The leafy cuttings need to be placed in a mist bed that provides intermittent misting. Once the cuttings are rooted, they can be planted outdoors.

Layering

Layering occurs in nature in a number of plants (e.g., forsythia, grape, and many brambles) where the flexible branches make contact with the soil. The primary use of layering in the fruit industry is in the production of clonal rootstocks. This is achieved either through mound layering or trench layering. In the former, which is used especially with apples, young plants of the rootstock grown in a stool bed are cut to within a few centimeters of the ground. When the new shoots appear, they are partially covered with soil, and this process is repeated during the growing season as the shoots continue their growth. At the end of the season, the mound of soil is removed and the rooted shoots harvested. The process is repeated in subsequent years, and well-maintained stool beds have a long life. The other method, referred to as trench layering, is essentially similar to mound layering, except in the initial step the plants of the rootstock are laid in a shallow trench and pegged in place. This forces the plants to generate new shoots, which are treated similar to those in mound layering.

Grafting

Grafting is the process of uniting two compatible plants—the scion (the cultivar) and the rootstock—to produce a single desirable plant. This is valuable from the standpoint of exploiting the benefits provided by clonal rootstocks or simply for vegetative propagation of a desirable scion (where a seedling rootstock is being used). With temperate fruit trees, grafting is done in winter or early spring, using dormant scion wood. Although many types of grafts may be used, one common type is the whip-and-tongue graft. In this, a diagonal cut is made through the rootstock, and then a vertical slit is made at the upper end of the rootstock. Matching cuts are made to the scion, and the two components are interlocked with each other, ensuring good cambium contact between the stock and the scion. The union is wrapped with airtight material to keep it from desiccating and to hold the two parts together during the healing process. A number of materials, including grafting tape and polyethylene strips, can be successfully used to wrap the union.

Budding

With budding, also referred to as bud grafting, a single bud from the scion is inserted into the rootstock, and this grows to produce the top of the tree. The advantages of budding over grafting are that it permits more efficient use of bud wood and allows for greater flexibility as regards time of conducting the operation. Although many types of bud-grafting techniques have been developed, two have been more widely used with temperate fruit: chip budding and shield budding (also referred to as T-budding).

Shield budding is conducted during the growing season when the cambium is active. A vertical incision about 2.5 centimeters long is made on the rootstock, followed by a horizontal second cut at the top of the first to produce a "T"-shaped incision. A bud is removed from the bud stick by making a cut starting about 1.3 centimeters below the bud and extending about 1.3 centimeters above the bud. For June and fall budding, the bud stick is obtained from an actively growing tree, so immediately after obtaining it, the leaves must be removed. The shield-shaped bud from the scion is inserted into the T-shaped cut in the rootstock, and cambium contact is readily achieved. The bark flap of the rootstock will cover the bud and keep it from desiccating, and tying the union with a rubber budding strip ensures that it is held securely until it has healed. Once the bud has formed a union with the rootstock, the top of the rootstock needs to be decapitated to allow the bud to grow.

Chip budding has a major advantage over shield budding in that it is not limited by the season and does not require that the cambium is actively dividing. Further, this method results in better graft union formation and is therefore widely used by commercial nurseries. The procedure requires making two cuts in the rootstock—a "down-stem" cut about 2.5 centimeters that is angled inward and a second, shorter, "up-stem" cut at an angle of approximately 45 degrees into the stock that meets with the first cut. This results in a triangular-shaped chip of bark and wood being removed from the rootstock. This is replaced by a similar-sized chip from the scion that contains a bud. Unlike a shield bud, a chip bud is not held in place or kept from desiccating by a bark flap, and the union must be wrapped with an airtight material.

Tissue Culture

Tissue culture (or micropropagation) provides a technique for rapid multiplication not only of herbaceous plants but also temperate fruit species. The procedure is relatively simple. Actively growing shoot tips, about 2.5 centimeters in length, are excised from the plant to be multiplied and implanted on sterile medium in a test tube or other appropriate container. The culture medium, usually solidified with agar, contains nutrients, sugar, and growth hormones. Supplementing the culture medium with a cytokinin (benzyladenine at 1 to 2 parts per million) induces the shoot tip to produce new shoots. These shoots can be excised and recultured on fresh medium at periodic intervals. Shoots can be rooted either in vitro (in culture medium supplemented with an auxin) or ex vitro (where they are treated as microcuttings). Rooted plantlets are propagated under conditions of high humidity and need to be gradually acclimated to ambient conditions.

Decisions made during the initial establishment of an orchard will have far-reaching ramifications. It is important that high-quality plants are obtained from a reputable nursery because they will have a significant impact on orchard operations and profitability.

Related Topics: CULTIVAR SELECTION; ROOTSTOCK SELECTION

SELECTED BIBLIOGRAPHY

Garner, R. J. (1976). *The grafter's handbook.* New York: Oxford Univ. Press.

Hartmann, H. T., D. E. Kester, F. T. Davies Jr., and R. Geneve (2002). *Plant propagation: Principles and practices.* Englewood Cliffs, NJ: Prentice-Hall.

Singha, S. (1986). Pear *(Pyrus communis).* In Bajaj, Y. P. S. (ed.), *Biotechnology in agriculture and forestry,* Volume 1 (Trees 1) (pp. 198-206). Berlin: Springer-Verlag.

Singha, S. (1990). Effectiveness of readily available adhesive tapes as grafting wraps. *HortScience* 25:579.

Zimmerman, R. H., R. J. Griesbach, F. A. Hammerschlag, and R. H. Lawson, eds. (1986). *Tissue culture as a plant production system for horticultural crops.* Dordrecht, the Netherlands: Martinus Nijhoff Publishers.

1. ROOTSTOCK SELECTION

– 1 –

Rootstock Selection

Curt R. Rom

A rootstock (understock, stock) is the root system of a grafted or budded plant. Most temperate zone fruit trees are propagated by the asexual methods of grafting or budding in order to preserve the characteristics of the aerial portion, or scion, of the plant. In some cases, the scion cultivar of the plant cannot be reproduced by seed or from adventitious roots on cuttings, and so propagation by grafting onto a rootstock is necessary. Rootstocks also are used for other purposes, such as tree size control, disease resistance, or winter hardiness.

Grafting was a horticultural art for several thousand years, and in the past three centuries, the potential was realized for using selected rootstocks to affect growth and performance of scions of plants. Early horticulturists initiated programs of selection, improvement, and hybridization for superior rootstock cultivars. Subsequently, pomologists interested in new fruit tree rootstocks established comparative trials and systematic experiments to determine edaphic, adaphic, and biotic adaptabilities and growth and productivity capabilities. This remains an important pursuit in horticulture.

EFFECTS OF ROOTSTOCKS AND REASONS FOR SELECTION

Rootstocks are used in tree fruit production for a number of reasons. The principle goal is to maintain specific characteristics of a fruit scion, since it does not propagate readily by other methods. A fruit cultivar does not come true from seed or may not form seeds, and cuttings made from the scion may not root readily. Further, if roots form on cuttings, they may not be environmentally adaptable, may have pest susceptibilities, or may confer unfavorable characteristics to the plant. Thus, a rootstock is used for the perpetuation, survival, and growth of the scion.

Rootstocks also may impart other characteristics to a scion cultivar. By virtue of width and depth of root growth extension, seasonal growth pattern, and root-wood fiber strength, rootstocks affect the anchorage and stability of the resultant grafted plant. Rootstocks may vary in tolerance to soil characteristics and thus be selected for adaptability to a specific soil's physical nature (texture, density, depth, and compaction), chemical nature (pH, salt content, cation exchange capacity), or environment (gas content, specifically oxygen and carbon dioxide, and water content). Rootstocks also express resistance or susceptibility to soil temperature extremes. Probably most limiting is cold temperature tolerance; therefore, rootstock hardiness is an important selection criterion in some growing regions. In addition, rootstocks express resistance or susceptibility to insect, nematode, disease, and vertebrate pests. All of these conditions—soils through pests—are considerations when selecting rootstocks.

The genetic expression of growth and adaptive variation to soil environment or pests notwithstanding, rootstocks cause changes in characteristics of the scion cultivar. A rootstock can control the genetic expression of scion growth (particularly the annual periodicity or season of growth), plant size, structural growth (angle of limbs), precocity (time from grafting until flowering), flowering date, and fruit maturity date. The basis for genetic control of a rootstock on a scion has not been clearly elucidated, although scientists have proposed models for phytohormone balance and interactions as well as assimilate (carbohydrate and nitrogen) feedback.

Whether it is a direct genetic expression or an indirect effect, rootstocks can also influence flower and fruit size and color, fruit firmness and flavor, mineral uptake and corresponding foliar and fruit mineral composition, scion cold hardiness, and floral frost tolerance. Moreover, changes in disease susceptibility are sometimes attributed to rootstocks.

ROOTSTOCK-SCION COMPATIBILITY

For a rootstock to be used, it must be graft compatible with the scion cultivar. Graft compatability is demonstrated as a successful union of rootstock and scion and is typified by flow of assimilates between the two parts, continued growth of the vascular transport tissues, and growth therefore of the whole plant. Generally, plants within a species are considered graft compatible, although there are incidences, albeit infrequent, of graft incompatibility between a rootstock and scion of the same species. Within some genera, there is widespread graft compatibility (e.g., *Prunus*). Thus, for crops such as peach, cherry, apricot, and almond, rootstocks may be a species other than the scion cultivar or may be hybrids among species. Compatibility may be even wider and occur among closely related genera within a family. For example, some graft compatibility for rootstocks has been demonstrated between pear *(Pyrus)* species, between pear and quince *(Cydonia),* and between pear and other Rosaceae genera.

TYPES OF ROOTSTOCKS

Rootstocks are broadly categorized into two groups based on how they are propagated: (1) seedling rootstocks and (2) clonal rootstocks (Table R1.1). Seedling rootstocks are propagated by gathering, stratifying, and then planting seed into a nursery. The genetic variability among seedling populations reflects the homozygosity of the species. For example, peach, which is relatively homozygous for most characteristics, will produce a seedling rootstock population that is, within reason, uniform. Thus, most peach rootstocks are commonly produced from seed. In contrast, seedling apples are highly variable within the seedling population for many characteristics, and apples are considered genetically heterozygous. Consequently, each seedling rootstock may grow and perform differently. To ensure uniformity of tree growth, most apple rootstocks are currently clones of a cultivar. Clonal rootstocks are propagated by asexual methods, and therefore within a rootstock population each individual is genetically similar to its population sibling. Clonal rootstocks are selected from naturally occurring populations by observation of rootstock performance, as developed hybrids from planned breeding programs, or as mutations of existing selections. Clonal rootstock propagation methods include layering, rooted cuttings, and micropropagation.

TABLE R1.1. Examples of tree fruit crops for which scion cultivars are commonly grafted onto rootstocks

Common Name	Botanical Name	Rootstocks Used	Examples
Apple	*Malus domestica*	Seedling	*M. domestica* spp.
		Clonal	M.7, M.26, M.9, MM.111, B.9, G.16
European pear	*Pyrus communis*	Seedling	*P. communis* 'Bartlett'
			P. betulaefolia
			P. calleryana
			Cydonia oblonga
		Clonal	*P. communis* 'Old Home'
			OHxF51, OHxF333
Peach/nectarine	*Prunus persica*	Seedling	*P. persica* 'Lovell'
			P. persica 'Bailey'
			P. persica 'Guardian'
			P. persica x *amygdalus*
Sweet cherry	*Prunus cerasus*	Seedling	*P. avium* 'Mazzard'
Tart cherry	*Prunus avium*		*P. mahaleb*
		Clonal	*P. avium* 'Mazzard F 12/1'
			P. avium 'Mazzard' x *mahaleb* (e.g., MxM2)
			P. cerasus
			P. avium x *pseudocerasus* 'Colt'
			Gisela 5, Gisela 6, Gisela 7

MULTICOMPONENT PLANTS

A rootstock and scion, combined by grafting or budding, results in a multicomponent plant—two different genetic systems mechanically combined into a single unit. This may be taken further, and additional pieces of plant material may be introduced into the plant system. For instance, if graft incompatibility exists between a rootstock and scion, an interstock (interstem) that is compatible to both is grafted between the two. Interstocks also are used to confer additional characteristics, such as size control, upon the plant system or to combine characteristics with the rootstock. For instance, a rootstock may be selected and used for its characteristics of anchorage and soil adaptability but may lack size control capacity. An interstock may be added to the plant system to confer size control and induce precocity.

In commercial fruit crop production, no single rootstock has proven to be perfect for all conditions and management systems. Continued development and experimentation with rootstocks and their propagation, production, and field use is needed. Thus, there should be a continued commitment to research on the biology and culture of rootstocks for tree fruit crops.

Related Topics: BREEDING AND MOLECULAR GENETICS; CULTIVAR SELECTION; DWARFING; PROPAGATION

SELECTED BIBLIOGRAPHY

Carlson, R. F. (1970). Rootstocks in relation to apple cultivars. In *North American apples: Varieties, rootstocks, outlook* (pp. 153-180). East Lansing, MI: Mich. State Univ. Press.

Cornell University New York State Agricultural Experiment Station (1998). *Geneva breeding programs.* Retrieved February 2, 2002, from <http://www.nysaes.cornell.edu/hort/breeders/index.html>.

Cummins, J. N. and H. S. Aldwinkle (1983). Breeding apple rootstocks. In Janick, J. (ed.), *Plant breeding reviews* (pp. 294-394). Westport, CN: AVI Publishing Co.

Hartmann, H. T., D. E. Kester, F. T. Davies Jr., and R. Geneve (2002). *Plant propagation: Principles and practices.* Englewood Cliffs, NJ: Prentice-Hall.

Hatton, R. (1917). Paradise apple stocks. *J. Royal Hort. Soc.* 42:361-399.

Rom, R. C. and R. R. Carlson (1987). *Rootstocks for fruit crops.* New York: J. Wiley and Sons, Inc.

Tukey, H. B. (1964). *Dwarfed fruit trees.* New York: The Macmillan Co.

Zeiger, D. and H. B. Tukey (1960). *An historical review of Malling apple rootstocks in America,* Bull. 226. East Lansing, MI: Mich. State Univ. Press.

S

1. *SOIL MANAGEMENT AND PLANT FERTILIZATION*
2. *SPRING FROST CONTROL*
3. *STORING AND HANDLING FRUIT*
4. *SUSTAINABLE ORCHARDING*

Soil Management and Plant Fertilization

Dariusz Swietlik

Nutritional needs of fruit trees are defined as the amounts of all mineral nutrients that must be acquired during one growing season to sustain normal growth, high productivity, and optimal fruit quality. Nutritional needs may be determined by seasonally excavating well-producing trees to quantify the amounts of nutrients absorbed, but this is a costly and labor-intensive task because trees have large canopies and extensive root systems. Furthermore, one must account for contributions of nutrient reserves, which are stored in old roots, trunks, branches, and twigs, toward nutrient demands of new tissues. The use of stable isotopes as nutrient tracers, e.g., ^{15}N, allows for distinguishing between the reserve and newly absorbed nutrients, thereby making such accounting possible.

A soil's natural store of nutrients may be sufficient to satisfy nutritional needs of fruit trees. Under most conditions, however, nitrogen (N), potassium (K), and calcium (Ca) and, less often, phosphorus (P), magnesium (Mg), boron (B), copper (Cu), iron (Fe), manganese (Mn), and/or zinc (Zn) must be supplied as fertilizers. Fertilizer needs of fruit trees are defined as the total amounts of all mineral nutrients that have to be applied as fertilizers to sustain normal growth, high productivity, and optimal fruit quality. Fertilizer and nutritional needs are numerically different because (1) the soil always satisfies at least a portion of plant needs for a given nutrient and (2) the efficiency of fertilizer recovery by plants is less than 100 percent due to losses caused by leaching, volatilization, or other chemical or biological interactions in the soil.

METHODS OF ESTIMATING FERTILIZER NEEDS

Plant Appearance

Fertilizer recommendations may be based on observations of plant leaves, shoots, and/or fruit, as these tissues produce characteristic symptoms of a given nutrient deficiency or excess. Generally, however, this method is inaccurate because (1) symptoms become visible only when nutritional maladies are severe; (2) many other factors, such as poor soil aeration, herbicide injury, insect damage, plant diseases, salinity, etc., may produce symptoms similar to those caused by nutritional disorders; and (3) visual plant observations are deceiving in cases of multielement deficiencies or excesses.

Soil Chemical Analysis

Soil samples, collected from the field, are extracted with a solution of weak acid, salt, chelating agent, or a combination of these compounds to determine the amount of a given nutrient, or its constant portion thereof, that will be released by the soil for plant uptake during the entire growing season. Results of such tests are expressed in kilograms per hectare or pounds per acre and are assigned a nutrient availability designator, e.g., low, medium, high, or very high.

Determining soil reaction or pH is the first, most important step in soil analysis, which then indicates the need for other tests. For example, a pH below 6.5 (acid soil) indicates that a lime requirement test should be performed, whereas pH above 7.0 (alkaline soil) indicates the need for salinity and exchangeable sodium tests. Soils are routinely tested for available K, P, and Mg, and less often for available B, Cu, Fe, Mn, and Zn.

Soil chemical tests have three major limitations. The first is the difficulty of obtaining a representative soil sample. This is due to high natural soil variability and the uncertainty of how many and where soil samples should be collected within the volume occupied by tree roots. The spatial distribution of tree roots and thus the extraction of nutrients from the soil profile are not uniform. There is, however, a lack of quantitative data on the pattern of nutrient extraction by the roots of fruit trees from different soil depths. Without this information, it is uncertain how sampling locations should be spatially distributed within the soil profile and how to properly interpret the results of soil tests conducted on samples collected from different soil depths. The second limitation reflects the fact that no chemical extracting procedure adequately mimics the natural process of nutrient release by the soil. The latter is affected by constantly changing biotic and abiotic conditions in the soil, i.e., the factors totally ignored by soil chemical tests. The third major limitation is the lack of sufficient data to correlate the results of soil tests with fruit tree responses in field fertilizer trials. It is well known that such responses may be modified by various production systems, rootstocks, and even fruit tree cultivars, thus further complicating the task of properly interpreting the results of soil tests.

Increasing sampling intensity, enhancing the knowledge of how the results of soil tests correlate with tree responses in field fertilizer trials, and the availability of trained personnel to interpret soil tests will mitigate the limitations discussed previously. Nevertheless, at the current stage of knowledge, the results of soil tests are only used to guide preplant fertilizer applications and in existing orchards to supplement the information obtained from tissue chemical analyses. Additionally, soil chemical tests are indispensable for properly managing extreme soil environments such as acidity, alkalinity, salinity, sodicity, and B toxicity.

Plant Tissue Analysis

Plant analysis is a widely accepted method of estimating fertilizer needs in deciduous orchards. Leaf chemical analysis is particularly useful in determining a fruit tree's nutritional status.

The results of leaf analysis are compared to leaf standards that designate sufficiency ranges of various nutrients. For the method to be reliable, sampling and handling of leaves must adhere to a prescribed protocol, which designates timing of collection, sampling pattern in an orchard, number of leaves per sample, size of a tree block represented by one sample, type of shoots from which the leaves should be sampled, and leaf washing, drying, and grinding procedures. Although the protocol may differ among various fruit-growing regions, generally, leaves (including petioles) are collected from the middle portion of current year, terminal shoots about 60 to 70 days after petal fall, i.e., from mid-July to mid-August for most of the fruit species grown in the Northern Hemisphere.

The concentration of a given element in a leaf integrates a number of factors that are known to affect the mineral nutrient status of the plant. These factors include, but are not limited to, soil nutrient availability, soil temperature and moisture, soil compaction and aeration, climatic conditions, plant genetics, rootstock, cultural and soil management practices, tree productivity and age, mechanical injuries, and damages caused by disease and arthropod organisms and nematodes. The multitude of these factors makes the interpretation of the results of leaf analysis difficult and requires the involvement of well-trained personnel.

The results of leaf analysis provide no information on causal factors that led to the development of a particular nutritional status. Also, they reflect past conditions that may or may not occur in the future. Frequent analyses, conducted every year or two, may overcome this limitation by revealing possible trends in nutritional status that can be related to culture, environment, soil, and other variables.

In some countries, preharvest analyses of apple fruitlets or postharvest analyses of apples are conducted to predict storage potential. These analyses usually include elements such as N, P, K, Ca, and Mg. The length of storage of a given lot of fruit is determined after comparing the results of analyses to specially developed standards for different apple cultivars.

Field Fertilizer Trials

Field fertilizer trials constitute the most reliable method of predicting fertilizer needs of fruit trees. This approach, however, requires the collection of multiyear data and employment of scientific staff and specialized equipment. Consequently, field trials are very expensive and are pursued only by research scientists. Unless correlated with leaf and/or soil analyses, the results of such trials cannot be easily extrapolated beyond the location of the study.

FERTILIZER APPLICATION PRACTICES

A fertilizer management program should start before an orchard is planted. At that time, fertilizers can be easily incorporated into the soil to the depth of 25 to 30 centimeters, i.e., the zone of high root activity. This is particularly important for P, K, Mg, and Ca fertilizers as well as Zn, Cu, Mn, and Fe, which move very slowly down the soil profile. When applied to the soil surface in mature orchards, these fertilizers will need a long time to reach the main part of the root system.

In the first two years after planting, fruit trees are usually fertilized individually by uniformly spreading fertilizers around each tree in a circle two to three times larger than the canopy. Depending upon local soil conditions, such applications may involve N and/or other fertilizers. In young, nonbearing orchards, N applications are often split into two or more applications that start in early spring prior to the beginning of current season growth and extend into midsummer. Later applications may unduly prolong vegetative growth, thus exposing trees to possible winter injuries.

In mature orchards, fertilizers are applied to weed-free strips within tree rows or are spread uniformly over the entire orchard floor. The recommended rates and timing of applications vary among fruit-growing regions and are influenced by the results of leaf and soil analyses, soil N mineralization rates, tree size, and orchard floor management practices. Typical N rates in apple orchards vary from zero to 100 kilograms per hectare. An early spring N application, prior to budbreak, stimulates current year growth but has a small impact on the N status of current season flower buds and fruit set. Such application, however, will have a more profound effect on the N status of flower buds and fruit set in the next growing season. Nitrogen applied in summer or early autumn will build up tree N reserves that will be utilized for flowering and fruit set the following spring. These late applications, however, may have deleterious effects on winter hardiness and/or fruit quality in the season of application.

Foliar fertilization with sprays of N, Ca, and microelements is practiced by fruit growers the world over. These sprays can supply nutrients directly to the foliage and fruit at times when they are most needed from the standpoint of tree productivity and/or fruit quality. For example, high fruit quality and long storage potential of apples and pears require high fruit Ca levels. Since these levels are usually higher than what a tree can supply from normal root uptake, pre-harvest foliar sprays or postharvest fruit dipping in a solution of Ca salts are widely practiced to enhance apple and pear quality and storage potential. Similarly, autumn foliar sprays with urea are practiced by some growers to elevate tree N reserves and fruit set the following spring. Under alkaline and neutral soil conditions, sprays with microelements, except Fe, are preferred over soil applications because of their enhanced effectiveness and a rapid plant response.

Fertigation, or applying nutrients with irrigation water, is another way fertilizers may be applied to fruit trees. This method is considered very efficient because it places nutrients in optimally wetted soil zones where roots are most active.

Conventional orchard management systems rely on the use of synthetic fertilizers to maximize orchard productivity and fruit quality. Sustainable orchard production systems, however, also emphasize orchard profitability, environmental protection, and soil conservation. Commonly recognized sustainable systems include integrated pest management (IPM), integrated fruit production (IFP), and organic production (OP). Under the IPM and IFP systems, soil conservation is emphasized and external inputs of synthetic fertilizers are minimized to achieve ecological stability and high profitability. Soil and/or plant chemical analyses are used to determine the need for fertilizer applications. Also, the methods, rates, and times of applications must be such as to minimize the risk of groundwater and surface water pollution. Mulching the soil strips beneath tree rows with organic composts is often practiced to supply plant nutrients, improve soil structure, and control weeds in order to minimize competition for water and nutrients. Under the OP system, no synthetic fertilizers are permitted, and soil fertility can be regulated only by applying natural composts or composted manures, or through the use of legume cover crops to enrich the soil in nitrogen and organic matter. Additionally, mineral nutrient sources that are mined or otherwise naturally occurring may be used, e.g., limestone to correct soil pH, calcium chloride for foliar sprays or postharvest dips to control physiological disorders on apples and pears, potassium sulfate, rock phosphate, copper hydroxide, etc. Leaf and soil analyses are pursued to verify the need for any particular element addition.

MANAGING EXTREME SOIL ENVIRONMENTS

Acid Soils

Soils are acid (pH below 7.0) due to chemical properties of the parent rock from which they were formed and/or because of the rainfall-induced leaching of soil cations, particularly Ca and Mg. The process is exacerbated by heavy N fertilization with acidifying fertilizers, e.g., ammonium sulfate, urea, or ammonium nitrate.

Deleterious effects of soil acidity include (1) aluminum (Al) and Mn toxicity to plants; (2) limited soil availability of P, Ca, and Mg; (3) reduced plant uptake of N and K; and (4) poor soil structure. Acid soils require liming to raise their pH to 7.0. The amount of lime can be determined by soil chemical analysis. When soil acidity is associated with low Mg availability, dolomitic lime is used to simultaneously enrich the soil in this element.

Alkaline Soils

Alkaline soils (pH above 7) are formed when evapotranspiration exceeds precipitation, as occurs under semiarid and arid climates, leading to the accumulation of cations in the soil, particularly Ca, Mg, and/or sodium (Na). Alkaline soils may also form in moderate climates when derived from rocks rich in calcium carbonate. Alkaline soils often contain free calcium carbonate, in which case they are called calcareous. Growing fruit trees on alkaline/calcareous soils poses serious challenges, such as deficiencies of Fe, Zn, Cu, and Mn, and low availability of P. Remedies include (1) foliar applications of Mn, Zn, and Cu; (2) soil applications of chelated forms of Fe; (3) sodding to increase soil Fe availability; and (4) judicious water management to avoid waterlogging, which is conducive to the development of Fe deficiency.

Salinity

Soil salinity develops when soils accumulate an excess of inorganic salts. This happens when evapotranspiration raises groundwater or perched water tables containing salts to the soil surface. Salinity may also develop when water containing high levels of soluble salts is used for irrigating crops. Remedial actions include (1) improving soil internal drainage and leaching the accumulated salts with an excess of irrigation water and/or (2) irrigating with water of acceptable quality. When the soil contains an excess of Na (sodic soil), the remedial actions include application of gypsum ($CaSO_4$) or, when the soil is calcareous, sulfur (S). This allows the excess of Na in the exchange complex to be replaced with Ca. Heavy irrigations to leach the excess Na must follow. Good internal soil drainage is a prerequisite for this measure to be successful.

As growers increasingly adopt environmentally friendly technologies to gain universal acceptance for the fruit they produce, mineral nutrient management will become more precise. The past practice of applying fertilizers as "an insurance policy" is no longer acceptable and has been replaced by practices based in sound science. With further advancements in estimating plant mineral nutrient requirements, acquisition, and use, new fertilization practices will supply nutrients only when needed to organs requiring them the most, and at optimal times from the standpoint of plant needs. Not only will this approach assure high productivity and quality of fruit, but it also will effectively protect the environment and conserve the soil.

Related Topics: ORCHARD FLOOR MANAGEMENT; ORCHARD PLANNING AND SITE PREPARATION; PLANT NUTRITION; SUSTAINABLE ORCHARDING

SELECTED BIBLIOGRAPHY

Miller, Raymond W. and Duane T. Gardiner (1998). *Soils in our environment.* Upper Saddle River, NJ: Prentice-Hall.

Organizing Committee of the International Symposium on Mineral Nutrition of Deciduous Fruit Crops (2002). World overview of important nutrition problems and how they are being addressed. Proc. Fourth Internat. Symposium. *HortTechnol.* 12:17-50.

Peterson, A. Brook and Robert G. Stevens (1994). *Tree fruit nutrition.* Yakima, WA: Good Fruit Grower.

Swezey, Sean L., Paul Vossen, Janet Caprile, and Wail Bentley (2000). *Organic apple production manual.* Oakland, CA: Univ. of California, Div. of Agric. and Nat. Resources.

Swietlik, Dariusz and Miklos Faust (1984). Foliar nutrition of fruit crops. *Hort. Rev.* 6:287-356.

– 2 –

Spring Frost Control

Katharine B. Perry

To practice frost protection successfully, one must understand the meteorology that creates freeze events. The sun's radiant energy warms the soil and trees in an orchard. When the soil and fruit trees become warmer than the air, they pass heat to the air by conduction. Through convective mixing, i.e., the warm air near the surface rising and being replaced by cooler air from above, hundreds of meters of the lower atmosphere are warmed. The soil and trees may also radiate heat into the atmosphere. Clouds and carbon dioxide can absorb or reflect some of this heat, trapping it near the surface. This is known as the greenhouse effect.

At night, with no incoming heat to warm the orchard, fruit trees lose heat through radiation and conduction until they are cooler than the surrounding air. The air then passes heat to the soil and trees, and the lower atmosphere cools. This creates an inversion, in which the temperature increases with altitude. The warmer air in the upper part of an inversion is an important source of heat for some frost protection methods.

FROST CONTROL PRINCIPLES

Although the terms "frost" and "freeze" are mistakenly interchanged, they describe two distinct phenomena. An advective, or wind-borne, freeze occurs when a cold air mass moves into an area, bringing freezing temperatures. Wind speeds are usually above 8 kilometers per hour, and clouds may be present. The thickness of the cold air layer ranges from 150 to 1,500 meters above the surface. The options for orchard protection under these conditions are very limited. A radiation frost occurs when a clear sky and calm winds (less than 8 kilometers per hour) allow an inversion to develop, and temperatures near the surface drop below freezing. The thickness of the inversion layer varies from 9 to 60 meters.

Other factors besides wind speed and clouds affect the minimum temperature that occurs. Growers in mountainous, hilly, or rolling terrain are familiar with frost pockets or cold spots. These are formed during radiation frosts by cold air drainage, i.e., cold, dense air flowing by gravity to the lowest areas of an orchard, where it collects. This causes temperatures to differ in relatively small areas, called microclimates.

Soil moisture and compaction can also have an effect on minimum temperature. A moist, compact soil will store more heat during the day than a loose, dry soil. Thus, it will have more heat to transfer to the trees at night. Groundcover reduces the heat stored in the soil below it. However, the frost protection disadvantages of groundcover management must be weighed against the benefits such as erosion control, dust reduction, improved soil aeration, etc.

FROST CONTROL TECHNIQUES

Site Selection

The best method of frost protection is good site selection. Choosing a site where frosts do not occur is optimum, but rarely possible.

Irrigation

During irrigation for frost protection, as 1 gram of water freezes, 80 calories of heat energy are released. As long as ice is being formed, this latent heat of fusion will provide heat. Irrigation for frost protection, also called sprinkler irrigation, can be accomplished by sprinklers mounted above or below the trees.

Although some risk is involved, the advantages of irrigation are significant. Operational cost is lower because water is much cheaper than oil or gas, and the system is convenient to operate because it is controlled at a central pump house. In addition, there are multiple uses for the same system, e.g., drought prevention, evaporative cooling, fertilizer application, and possibly pest control.

There are also disadvantages. The most important is that if the irrigation rate is not adequate, the damage incurred will be more severe than if no protection had been provided. Inadequate irrigation rate means that too little water is being applied to freeze at a rate that will provide enough heat to protect the crop. The situation is made complex by another property of water—evaporative cooling, or the latent heat of evaporation. As 1 gram of water evaporates, 600 calories of heat energy are absorbed from the surrounding environment. When compared to the 80 calories released by freezing, one sees that more than seven times more water must be freezing than evaporating to provide a net heating effect. An ice-covered plant will cool below the temperature of a comparable dry plant if freezing stops and evaporation begins. Since wind promotes evaporative cooling, wind speeds above 8 kilometers per hour limit the success of irrigation for frost protection. Further, with overhead irrigation, ice buildup can cause limb breakage, soils can become waterlogged, and nutrients can leach out. Also, most systems have a fixed-rate design; i.e., the irrigation rate cannot be varied. This means systems are designed for the most severe conditions and so apply excess water in most frosts.

Fog

Artificial fog is based on the greenhouse effect. If a "cloud" can be produced to cover the orchard, it decreases the radiative cooling. There has been some experimental success, but a practical system has not been developed. The difficulties lie in producing droplets large enough to block the outgoing long wave radiation and in keeping them in the atmosphere without losing them to evaporation.

Wind Machines

Wind machines capitalize on the inversion development in a radiation frost. They circulate the warmer air above down into the orchard. Wind machines are not effective in an advective freeze. A single wind machine can protect approximately 4 hectares, if the area is relatively flat and round. A typical wind machine is a large fan about 5 meters in diameter mounted on a 9-meter steel tower.

Wind machines use only 5 to 10 percent of the energy per hour required by heaters. The original installation cost is quite similar to that for a pipeline heater system, making wind machines an attractive frost control alternative. However, they will not provide protection under windy conditions. Wind machines are sometimes used in conjunction with heaters. This combination is more energy efficient than heaters alone and reduces the risk of depending solely on wind machines. When these two methods are combined, the required number of heaters per hectare is reduced by about half.

Helicopters have been used as wind machines. They hover in one spot until the temperature is increased enough and then move to the next area. Repeated visits to the same location are usually required.

Heaters

Heating for protection has been relied upon for centuries. The increased cost of fuel has diminished the popularity of this method; however, there are several advantages to using heaters that alternatives do not provide. Most heaters are designed to burn diesel fuel and are placed as free-standing units or connected by a pipeline network throughout the orchard. Advantages of connected heaters are the abilities to control the rate of burning and to shut all heaters down from a central pumping station simply by adjusting pump pressure. A pipeline system can also be designed to use natural gas.

Heaters provide protection by three mechanisms. The hot gases emitted from the top of the stacks initiate convective mixing in the orchard, tapping the important warm air source above in the inversion. About 75 percent of a heater's energy is released in this form. Most of the remaining 25 percent of the total energy is released by radiation from the hot metal stack. This heat is not affected by wind and will reach any solid object not blocked by another solid object. Heaters may thus provide some protection under wind-borne freeze conditions. A relatively insignificant amount of heat is also conducted from heaters to the soil.

Heaters provide the option of delaying protection measures if the temperature unexpectedly levels off or drops more slowly than predicted. The initial installation costs are lower than those of other systems, although the expensive fuels required increase the operating costs. There is no added risk to the crop if the burn rate is inadequate; whatever heat is provided will be beneficial.

Chemicals

The objective of having an inexpensive frost protection material that can be stored until needed and easily applied has existed since the mid-1950s. Numerous materials have been examined. These fall into several categories but, in general, they have been materials that allegedly (1) change the freezing point of the plant tissue, (2) reduce the ice-nucleating bacteria on the crop and thereby inhibit ice/frost formation, (3) affect growth, e.g., delay dehardening, or (4) work by some "undetermined mode of action." To this author's knowledge, no commercially available material has successfully withstood the scrutiny of a scientific test. However, several products are advertised as frost protection materials. Growers should be very careful about accepting the promotional claims about these products. Research continues, and some materials have shown some positive effects.

The proper method of frost protection must be chosen by each grower for the particular site considered. Once the decision has been made, several general suggestions apply to all systems. If frost protection is to be practiced successfully, it must be handled with the same care and attention as spraying, fertilizing, pruning, and other cultural practices. Success depends on proper equipment used correctly, sound judgment, attention to detail, and commitment. Growers should not delegate protection of the crop to someone who has no direct interest in the result. It is important to prepare and test the system well before the frost season begins, to double-check the system shortly before an expected frost, and not to take down the system before the threat of frost has definitely passed. Problems that are handled easily during the warm daylight can become monumental and even disastrous during a cold, frosty night when every second counts.

Related Topics: DORMANCY AND ACCLIMATION; FLOWER BUD FORMATION, POLLINATION, AND FRUIT SET; TEMPERATURE RELATIONS

SELECTED BIBLIOGRAPHY

Barfield, B. J. and J. F. Gerber, eds. (1979). *Modification of the aerial environment of plants.* St. Joseph, MI: ASAE Monograph.

Hoffman, G. J., T. A. Howell, and K. H. Solomon, eds. (1990). *Management of farm irrigation systems.* St. Joseph, MI: ASAE Monograph.

Perry, K. B. (1998). Basics of frost and freeze protection for horticultural crops. *HortTechnol.* 8:10-15.

Rieger, M. (1989). Freeze protection for horticultural crops. *Hort. Rev.* 11:45-109.

Rosenberg, N. J., B. L. Blad, and S. B. Verma (1983). *Microclimate: The biological environment.* New York: John Wiley and Sons.

– 3 –

Storing and Handling Fruit

A. Nathan Reed

Once fruit are harvested, the main goal is to maintain freshness and quality. Cooling is the primary mechanism used to minimize reduction in quality. A fruit is a living entity and continues to metabolize following harvest. Respiration is a major part of metabolism and is the process of breaking down stored carbohydrates to produce energy. A warm fruit has a higher rate of respiration, which leads to accelerated ripening, depleted energy reserves, and decreased potential storage life. The faster a fruit respires, the quicker it will ripen and eventually deteriorate. For example, lowering the temperature of 'Granny Smith' apples from 20 to 0°C decreases the rate of deterioration by a factor of five. Exposing fruit to direct sunlight can lead to elevated respiration rates and internal fruit temperatures that greatly exceed the surrounding ambient temperature. Removing fruit from direct sunlight is the first step in the cooling process. Placing fruit in the shade or using mechanical covers and reflective materials can significantly decrease surface and internal fruit temperatures. Lowering fruit temperature by 10°C reduces respiration rate by a factor of two and also reduces reactivity to ethylene, a gaseous ripening hormone.

PRECOOLING

Precooling is a technique used to lower the temperature of fruit prior to cold storage or shipping. Precooling began in the early 1900s to prepare fruit for rail shipment or export. The importance of precooling depends on the fragility of the product and the storage life expectancy. The choice of the precooling temperature depends on temperature at harvest, sensitivity of the fruit to chilling, physiology of the product, and potential postharvest storage life. The rate of cooling depends on various factors: (1) rate of heat transfer from the fruit to the surrounding medium, (2) temperature difference between the fruit and the cooling medium, (3) nature and physics of the cooling medium, and (4) thermal conductivity of the fruit. Various methods to cool harvested fruit fall into passive or active categories. Passive techniques include shading, placing fruit in a cold room, or simply adjusting harvest time to early or late in the day to avoid the highest midday temperatures. Active techniques include hydrocooling or forced-air cooling. Most tree fruit are cooled by the cold-air method. However, stone fruit can be rapidly and effectively cooled by hydrocooling, which involves a cascading cold-water drench.

Hydrocooling requires the most handling but is the quickest form of heat removal. The cooling medium, water, comes in direct contact with the entire fruit surface, and heat is given up to the liquid water, which has a higher heat transfer coefficient than air. The target temperature for water in a hydrocooler is 0°C. The rate at which hydrocooling will lower fruit temperature depends on several factors: (1) temperature differential between the fruit body and the water, (2) velocity of water cascading over the fruit, (3) relative surface area being covered by the cool water, (4) thermal conductivity of the water, and (5) total volume of water being used. Active versus passive system efficiency is illustrated in peaches by the time required to lower the temperature by 90 percent of the difference between the initial fruit temperature and the cooling medium temperature. Hydrocooling can achieve the temperature goal in approximately 20 minutes, while forced-air cooling requires three hours, and passive room cooling requires 18 hours. In addition to handling considerations, hydrocooling involves potential contact with liquid-carrying pathogenic spores from other infected fruit and an increased risk for decay. Sanitation of the water drench is important and can be managed through the use of products such as chlorine or chlorine dioxide. Monitoring and maintaining these disinfectants at effective levels is important to minimize decay losses.

Some temperate fruit do not lend themselves to the direct water contact of hydrocooling techniques, and thus air cooling must be used to remove field heat. For forced-air cooling to be effective, careful planning is essential. Refrigeration capacity, container design and placement, along with fan placement, size, architecture, and speed, govern the rate and efficiency of cooling. The goal of forced-air cooling is to keep cold air moving at 61 to 122 meters per minute across the fruit. Rapid cooling in this fashion relies on maximum air movement around individual fruit rather than around containers. Large containers can significantly reduce the rate of air penetration and therefore increase cooling time for fruit that comprise the center core of a bin. Vented bulk containers used in forced-air cooling should allow rapid airflow and disbursement of the flow to come in contact with as much fruit surface area as possible. Tight stacking of bins in cold rooms is essential for maximizing cooling rates. Without tight stacking or with misalignment of bins, boxes, or crates, airflow patterns can be disrupted and result in short cycling and some produce not being cooled actively. Bin or carton stacking into the configuration of a tunnel, with fans placed directly against the container stacks, provides a configuration that promotes a negative pressure gradient and draws rather than pushes cold air across and around fruit from spaces between and vents within containers. Tarps can be used to seal openings that otherwise would allow the air to short-circuit and not come in contact with fruit. Serpentine cooling is another form of forced-air cooling in which barriers are placed over alternating forklift openings, thus forcing air to move either up or down through the fruit as it moves toward the fans.

COLD STORAGE

Storage and precooling of produce in a cold room are two distinctly different processes. Once fruit have been precooled, they should remain in cold storage to maintain quality and increase storage life. Cold-temperature storage can be attained through the use of a mechanical refrigeration system that works on the principle of a liquid absorbing heat as it changes states to a gas. Most commonly, a halide refrigerant or ammonia is compressed into liquid form and then released to a gas state under the control of an expansion valve. A series of coils within the refrigerated storage area serves as the mechanism in which this expansion and absorption of heat occurs. As the liquid refrigerant evaporates to a gas within the coils, heat from the fruit is absorbed. Within the closed system, the heated vapor refrigerant is compressed, and a process of condensation removes heat. The cooled refrigerant is compressed to liquid form, and the process repeats itself. The process of maintaining cold-temperature storage requires less horsepower and uses less energy than precooling. Research shows that cycling fans within a cold-storage room or using variable frequency drives can reduce the total energy consumed by motors without affecting fruit quality.

Storing temperate fruit requires careful coordination with respect to temperature (Table S3.1). It is important to maintain specific temperatures that do not drift appreciably above or below set limits. Being off by a degree for an extended period of time can have serious ramifications for the potential storage life and the commercial viability of the fruit itself. In most cases, it is better to be slightly high in the target range than slightly low. Peach storage operators are specifically warned to avoid the temperature range of 2 to 8°C, as peaches stored within this range can develop a serious malady termed "woolliness." Most apples are stored close to 0°C. Dense apples such as 'Braeburn', 'Fuji', and 'Granny Smith' are sensitive to internal carbon dioxide buildup and are usually stored at elevated temperatures of 2 to 3°C.

TABLE S3.1. Physical properties and storage considerations for temperate fruit

	% Water Content	Highest Freezing Point (°C)	Storage Temperature (°C)	% Relative Humidity in Storage	Approximate Storage Life
Apples	84.1	−1.5	−1.0 - 4.0	90-95	1-12 months
Apricots	85.4	−1.0	−0.5 - 0.0	90-95	1-3 weeks
Cherries, sour	83.7	−1.7	0.0	90-95	3-7 days
Cherries, sweet	80.4	−1.8	−1.0 - −0.5	90-95	2-3 weeks
Nectarines	81.8	−0.9	−0.5 - 0.0	90-95	2-4 weeks
Peaches	89.1	−0.9	−0.5 - 0.0	90-95	2-4 weeks
Pears	83.2	−1.5	−1.5 - 0.5	90-95	2-7 months
Plums	86.6	−0.8	−0.5 - 0.0	90-95	2-5 weeks

Source: Hardenburg, Watada, and Wang, 1986.

CONTROLLED ATMOSPHERE STORAGE

Altering the atmospheric gas content of the refrigerated room in which fruit are stored can lengthen the effective duration of storage time. The concept of controlled atmosphere (CA) storage originated with work on modified atmospheres around 1920. Research on low-oxygen storage did not begin until around 1930. The initial use of low-oxygen CA storage did not occur until the 1940s. Several temperate fruit respond well to CA storage. Excellent benefits are obtained with apples and pears, while good effects are obtained with cherries, figs, nectarines, peaches, plums, and prunes. Fair results are obtained with apricots. The overall impact of CA storage is to allow retailers to provide specific cultivars over a longer marketing period, thus removing seasonal fluctuation in product availability. Physiological effects of CA on fruit quality include lowering respiration, reducing acidity and chlorophyll losses, and maintaining firmness. A negative effect of long-term CA storage is a reduction in the ability of fruit to produce characteristic flavor and volatile components.

Controlling storage gas levels requires an airtight room and can be accomplished through a number of techniques. The removal of oxygen from an airtight storage room can be accomplished by flushing the room with compressed nitrogen, or in a more expensive procedure, nitrogen can come from the release of the gas from liquid nitrogen as it warms and boils. One of the first methods developed to reduce the oxygen content of ambient air, from roughly 21 to about 3 percent, used a catalytic burner. Another method for producing nitrogen gas is labeled "cracking." In this process, a burner cracks ammonia into nitrogen and hydrogen. The hydrogen is burned, producing water as a by-product. More recently, the methods for CA atmospheric generation rely on gas separation technology. One technique, pressure swing adsorption, utilizes two chambers filled with carbon molecular sieve material that retains oxygen and generates relatively pure nitrogen. As oxygen is adsorbed, the pressure builds on one chamber, and nitrogen is released into the CA room. As one chamber absorbs oxygen, the other chamber is desorbing. The carbon sieve also acts as a scrubber that removes carbon dioxide and ethylene. Another gas separation technique uses ambient atmosphere and separates nitrogen from oxygen through the use of a hollow fiber, molecular sieve membrane system. This system was first developed in the medical field in the 1950s for use in artificial kidneys. For CA purposes, ambient air is compressed, put through a coalescing filter, dried, and filtered for particles, oil contaminants, and carbon-containing materials. Once cleaned and filtered, the air passes through a bundle of hollow fibers. Oxygen, carbon dioxide, and water vapor are "fast" gases that diffuse through the membrane. Nitrogen is slow and thus remains behind within the membrane bundle at purity levels around 98 percent. Some CA atmosphere regimes require oxygen levels of 0.7 percent, and these low levels can be achieved through recirculating the 98 percent pure nitrogen through the hollow fiber sieve a second time.

In addition to regulating oxygen levels, the buildup of carbon dioxide in storage is a concern. Carbon dioxide in the past was mainly controlled by the use of alkaline solutions such as potassium or calcium hydroxide, an activated charcoal scrubber, or adsorption by dry hydrated lime. Hydrated lime is a method still used today. The use of gas separators allows the control of carbon dioxide by dilution with liquid or compressed nitrogen.

RELATIVE HUMIDITY

Relative humidity (RH) is another important consideration in storing fruit and is defined as a ratio of the quantity of water vapor present to the potential maximum amount of vapor at a given temperature and pressure. The major constituent in a piece of fruit is water (Table S3.1). Intercellular space within healthy fruit is saturated at approximately 100 percent RH. To decrease the exposure time requirement in precooling, refrigeration coils have relatively low set points, which create a large difference between coil and fruit temperature. Actively growing and harvested fruit regulate temperature and give up heat by releasing water vapor. The cooling process results in an unavoidable loss in fruit weight. The relatively large amount of water given up by fruit during the precooling process is frozen on the coils and requires several defrost cycles each day. After fruit reach the targeted storage temperature, the objective in the cold-storage room changes to maintaining a stable storage temperature. At this point, coil temperature can be raised to roughly 0.5°C lower than the targeted fruit storage temperature.

A refrigerated storage room filled with fruit will come to equilibrium at around 85 to 90 percent RH. As air circulates around the cold coils, suspended water vapor freezes on the coils and the RH of the room is lowered. The difference between the internal fruit RH and the surrounding atmosphere creates a gradient that draws additional moisture out of fruit. Several techniques can be employed to reduce fruit water loss. Restricting fan speed to lower velocities of 15 to 23 meters per minute can reduce the rate of water loss. Covering individual fruit bins or containers with polyethylene increases the RH of the atmosphere surrounding fruit and slows the rate of water loss. Compared to conventional wooden bins that can absorb water equivalent to approximately 15 percent of their weight, plastic bins do not absorb water and thus reduce fruit shrinkage due to water loss. Adding standing water to the floor of a cold-storage room does little to raise the RH or prevent fruit weight loss, as water changes physical states from liquid to vapor at the interface of the liquid and the atmosphere. The best method to increase RH is to supply water in a fashion that increases the total available surface area of the liquid. This is best accomplished by adding water in the form of a fine fog. The combined surface area of a fog comprised of millions of small, micron-sized droplets is several orders of magnitude greater than the entire surface area of a floor covered with standing water.

With humidity, relative is an important term. In contrast to warm air, cold air has a very limited capacity to support water vapor. The atmosphere of a 100 cubic meter cold room at 0°C and 100 percent RH can support 480 grams of water. The difference between 85 and 95 percent RH at 0°C storage can have significant quality effects, even though it represents only 50 grams of suspended water. Keeping fruit at or close to 100 percent RH over long periods of time, however, can have some detrimental effects, such as fruit splitting, skin cracking, or increased decay. Extremely hydrated fruit also are more susceptible to bruising during handling and packing following storage.

MONITORING STORAGES

Cold-storage and CA facilities are monitored for temperature, RH, ammonia, oxygen, and carbon dioxide. Refrigeration equipment for cooling and maintaining fruit temperature should have reliable control mechanisms. Consistent, stable temperatures maximize storage life. Temperatures outside suggested ranges reduce expected storage life if too high or can lead to severe chilling or freeze injury if too low. An obstacle in managing RH is the ability to measure it accurately. Sensors for measuring RH include capacitive, resistive, thermal conductivity, and dew point instruments. The most reliable sensors are chilled mirror hygrometers that measure the dew point.

Specific cultivars of fruit have specific atmospheric requirements. Storage at a gas concentration that is 0.5 percent lower or higher than recommended can result in a commercial disaster. Some cultivars are very sensitive to carbon dioxide at concentrations above 0.5 percent. Thus, it is important to be able to monitor gas levels on a routine schedule, automatically control these gas levels within set points, and have storage operators on call 24 hours a day for emergencies. The sensors used for detecting oxygen are paramagnetic, polarographic, or electrochemical. Carbon dioxide sensors are normally infrared detectors. The latest technology of scanning near-infrared equipment has demonstrated the ability to monitor carbon dioxide and other important gases such as ethylene, carbon monoxide, propane, 1-methyl-cyclopropene, and various flavor components inside storage rooms. This equipment is quite expensive, but with a manifold system of solenoid valves and pumps, one scanning infrared sensor can monitor several rooms on a routine cycling basis.

The main objective in handling and storing freshly harvested fruit is to preserve quality through rapid cooling and cold storage. The most valuable tools for storage operators are accurate, reliable, and stable equipment that produce, maintain, and monitor cold-temperature and atmospheric conditions.

Related Topics: FRUIT MATURITY; HARVEST; MARKETING; PACKING; PHYSIOLOGICAL DISORDERS; POSTHARVEST FRUIT PHYSIOLOGY

SELECTED BIBLIOGRAPHY

Combrink, J. C. (1996). *Integrated management of post harvest quality.* Stellenbosch, South Africa: Infruitec.

Eskin, N. A. Michael, ed. (1991). *Quality and preservation of fruits.* Boca Raton, FL: CRC Press.

Hardenburg, R. E., A. E. Watada, and C. Y. Wang (1986). *The commercial storage of fruits, vegetables, and florist and nursery stocks,* Handbook No. 66. Washington, DC: USDA.

Kader, A. (2002). *Postharvest technology of horticultural crops,* Pub. 3311. Davis, CA: Univ. of California, Coop. Exten. Serv., Div. of Agric. and Nat. Resources.

Kays, S. J. (1997). *Postharvest physiology of perishable plant products.* Athens, GA: Exon Press.

LaRue, J. H. and R. S. Johnson (1989). *Peaches, plums and nectarines: Growing and handling for fresh market,* Pub. 3331. Davis, CA: Univ. of California, Coop. Exten. Serv., Div. of Agric. and Nat. Resources.

Little, C. R. and R. J. Holmes (2000). *Storage technology for apples and pears.* Knoxfield, Victoria, Australia: Dept. of Nat. Resources and Envir.

Wills, R., B. McGlasson, D. Graham, and D. Joyce (1998). *Postharvest: An introduction to the physiology and handling of fruit, vegetables and ornamentals,* Fourth edition. Adelaide, South Australia: Hyde Park Press.

Sustainable Orcharding

Tracy C. Leskey

Conventional orchard management is guided by the goal of maximizing bearing potential per hectare in order to increase short-term gains. Within this framework, growers typically rely on management practices that are linked to external or off-farm inputs. These external inputs include synthetic pesticides used to control insects, diseases, and weeds; synthetic fertilizers and irrigation systems; and synthetic growth regulators used to control numerous aspects of fruit production, such as budbreak or bloom, fruit set, preharvest drop, size, and color. Shortcomings of reliance on these inputs include pesticide resistance, soil degradation, collateral injury to nontarget organisms, and concerns for human health. Given their reliance on off-farm inputs to establish and maintain production, these management systems reduce long-term sustainability despite increases in short-term gains. Sustainable production in agricultural systems must include consideration of economics and profitability, environmental protection, conservation of natural resources, and social responsibility (Reganold, Papendick, and Parr, 1990). Alternative production systems that view an orchard as a potentially sustainable agroecosystem are becoming more widely accepted as management strategies are developed that lead to less reliance on external inputs. Three widely studied approaches to achieving sustainable orchard production are integrated pest management, integrated fruit production, and organic production.

INTEGRATED PEST MANAGEMENT

Integrated pest management (IPM) has been characterized as a decision-based process that involves coordination of multiple tactics for optimal control of all classes of pests (insect, disease, weed, and vertebrate) in an economically and environmentally sound manner (Prokopy, 1993), thus leading to a greater level of sustainability of the system. There are two approaches to implementing IPM in deciduous tree fruit production. The first begins with a conventionally managed orchard that transitions to a more economically and environmentally sustainable agroecosystem as external inputs are reduced or eliminated in a stepwise manner. Alternatively, a second approach establishes a more natural orchard ecosystem in which external inputs are used only to augment natural processes (Brown, 1999).

IPM programs emphasize prevention, encouraging growers to choose sites, rootstocks, cultivars, and planting systems that will lead to ecological stability and economic viability. Soil conservation is also emphasized. A multiple tactic approach for pest control is one of the key components of IPM programs. For example, herbicide applications are not the sole basis for weed control. Instead, mechanical control methods such as mowing and tillage and cultural methods such as mulching are supplemented by herbicides. Monitoring is another key element of any IPM program, allowing growers to determine if and when pesticide applications are required for control of a particular pest. This approach is different from conventional systems that traditionally relied on residual activity as a guide to determine when subsequent applications were necessary. However, there is variability among IPM production systems as to the acceptability of particular management practices, as highlighted by the differences in the two fundamental approaches to IPM described earlier. For example, differences exist as to whether chemical thinning agents, synthetic plant growth regulators, and postharvest fungicide treatments can be used. Thus, the level of sustainability of IPM production systems, though considered to be greater than conventional systems, likely varies among practitioners.

Varying degrees of either IPM approach are commonly practiced in tree fruit production areas throughout the world. Until recently, there were no certification and/or marketing programs supporting IPM production. Therefore, it was difficult to justify the higher market price of fruit grown under IPM regimes. However, with certification and labeling programs, and other growing and marketing programs either under development or in practice, more tree fruit are being grown and marketed to reflect both the environmental and economic considerations of IPM programs.

INTEGRATED FRUIT PRODUCTION

Integrated fruit production (IFP) is defined as "economical production of high quality fruit, giving priority to ecologically safer methods, minimizing the undesirable side effects and use of pesticides, to enhance the safeguards to the environment and human health" (Cross and Dickler, 1994, p. 2). IFP systems emphasize long-term sustainability by minimizing the use of external inputs into the agroecosystem.

There are many differences between conventional and IFP production systems. For example, in new orchards, IFP programs encourage growers to select sites, rootstocks, cultivars, and planting systems that will require minimal external inputs but lead to both ecological stability and economic success. Soil sterilization generally is not permitted or is highly discouraged, and planting spaces are required or encouraged to be large enough to accommodate trees throughout their life span without the use of synthetic plant growth regulators and/or severe pruning. IFP programs require that growers practice soil conservation by recycling organic matter when possible and apply fertilizers only if results of soil and/or plant chemical analysis identify a specific nutrient deficiency. When deemed necessary, fertilizers are to be applied in a manner to minimize risk to groundwater, and to minimize soil compaction and/or erosion. IFP programs encourage pruning practices and training systems to achieve balanced and sustained productivity and quality. Fruit thinning by hand is recommended and preferred, although chemical thinning agents are permitted on some cultivars. Many IFP programs do not permit synthetic plant growth regulators for improving fruit finish and color or to regulate ripening. Under IFP, multiple tactic management of insects, diseases, and weeds is given priority over synthetic pesticides. Conservation of natural enemies such as predatory mites and parasitic wasps is required. Bare soil management and use of residual herbicides generally are not recommended or are not permitted. Therefore, weed-free strips beneath fruit trees are to be maintained by cultural means such as mulching or mechanical cultivation, with herbicides acting as a supplemental treatment only. Monitoring is an essential component of pest management for IFP production. Any pesticide that is used in IFP systems must be one that is approved under IFP guidelines. These particular pesticides are considered to be the most selective, least toxic, and least persistent and are to be used at the lowest effective rates. Under IFP, fruit are harvested according to cultivar maturity, and if stored, then done so in a manner to maintain internal and external quality. Depending on the region of an organization-specific IFP program, postharvest treatment with some synthetic antioxidants and fungicides may not be permitted.

This particular production system is currently in use in the United States, Europe, and other fruit-growing regions throughout the world. European programs require that growers be professionally trained in IFP-endorsed management practices. Growers must attend an introductory training session as well as periodic updates. To become certified in Europe, a grower must practice at least a minimum number of IFP-endorsed guidelines for fruit production and must maintain accurate records of these practices and make them available for inspection by IFP administrators. In the United States, certification may or may not be part of the program; some, but not all, growers using a singular IFP plan to produce tree fruit undergo certification (Hood River District Integrated Fruit Production Program, 1997; Core Values Northeast, 2001). Regardless, these programs and production systems found in the United States, Europe, and other fruit-growing regions throughout the world may use an IFP label to market to consumers that fruit have been grown with both economic and environmental considerations.

ORGANIC PRODUCTION

Organic production (OP) has been defined as a "holistic production management system that promotes and enhances agroecosystem health, including biodiversity, biological cycles, and soil biological activity" (FAO/WHO Codex Alimentarius Commission, 2001, p. 3). This production system emphasizes use of cultural, mechanical, and biological management practices instead of external inputs such as synthetic pesticides. Production is based on management practices for site-specific conditions that enhance the ecological balance of natural systems (U.S. Department of Agriculture, Agriculture Marketing Service, 2001).

In a recent study, Reganold et al. (2001) found, in comparisons of conventional, integrated, and OP systems in apple, organic management achieved the greatest levels of both economic and environmental sustainability, followed by integrated, and then by conventional systems. Because of the minimal use of synthetic inputs and greater emphasis placed on long-term sustainability, there are many differences between conventional and OP systems. One of the greatest differences between conventional and organic systems is the emphasis placed on soil health and conservation in OP systems. Organic systems utilize compost, manure, or other natural organic matter to improve soil fertility, which will in turn be used by fruit trees, while conventional systems rely on synthetic fertilizer applications to supply nutrients to fruit trees themselves and not necessarily to improve overall soil fertility or sustainability. The specific nutrient contribution of composted organic matter can vary, and therefore soil and nutrient management requires careful, long-term planning (Edwards, 1998). Organic growers must carefully consider orchard sites, rootstocks, and cultivars. Rootstock as well as cultivar choices are extremely important, as some are more resistant to specific pests and diseases. Some regions may not be amendable for OP. For example, in areas of high rainfall, control of moisture-dependent diseases may not be economically feasible under OP standards because synthetic pesticides and genetically engineered plants are not permitted. As with IPM and IFP, cultural controls such as cover crops, cultivation, or mulches are used to manage weeds. However, unlike IFP, IPM, or conventional systems, no synthetic inputs, e.g., herbicides, are permitted. Therefore, weed control is a major challenge for OP, especially in young orchards where weed competition can reduce bearing potential. If a control strategy is needed for insect pests, growers utilize augmentation of natural enemies, mating disruption, traps, and barriers. Pests and diseases also can be controlled with botanical or other naturally occurring pesticides that are approved under the guidelines for OP certification programs. However, these materials can have adverse effects on nontarget organisms. Therefore, components of IPM programs, such as monitoring to determine when applications of these materials are necessary, are important in OP as well. Under OP, only hand thinning is permitted.

Because postharvest treatments with synthetic fungicides are not permitted, cultural practices such as sanitation in orchards and packinghouses are crucial for reduction of fruit rot and contamination if fruit are held in storage. Currently, growers who meet OP criteria can be certified under local, regional, or national programs to use an organic label to market their fruit as grown under what is considered to be an ecologically responsible production system based on long-term sustainability.

Public demand for high-quality fruit grown with ecological and environmental considerations is likely to continue. Therefore, emphasis on development and implementation of alternative production systems such as IPM, IFP, and OP also will increase.

Related Topics: DISEASES; INSECTS AND MITES; ORCHARD FLOOR MANAGEMENT; PLANT-PEST RELATIONSHIPS AND THE ORCHARD ECOSYSTEM; SOIL MANAGEMENT AND PLANT FERTILIZATION

SELECTED BIBLIOGRAPHY

Blommers, L. H. M. (1994). Integrated pest management in European apple orchards. *Annu. Rev. Entomol.* 39:213-241.

Brown, M. W. (1999). Applying principles of community ecology to pest management in orchards. *Agric., Ecosystems and the Environ.* 73:103-106.

Core Values Northeast (2001). Retrieved October 10, 2001 from <http://www.corevalues.org>.

Cross, J. V. and E. Dickler (1994). Guidelines for integrated production of pome fruits. *IOBC Tech. Guideline* III 17:1-12.

Edwards, L. (1998). *Organic tree fruit management.* Keremeos, British Columbia: Certified Organic Associations of British Columbia.

FAO/WHO Codex Alimentarius Commission (2001). *Guidelines for the production, processing, labeling and marketing of organically processed foods,* Rev. 1, GL 32-1999. Rome, Italy: FAO/WHO.

Hood River District Integrated Fruit Production Program (1997). Retrieved October 9, 2001 from <http://www.orst.edu/dept.hort/orchardnet/hifp.htm>.

Prokopy, R. J. (1993). Stepwise progress toward IPM and sustainable agriculture. *The IPM Practitioner* 15:1-4.

Reganold, J. P., J. D. Glover, P. K. Andrews, and H. R. Hinman (2001). Sustainability of three apple production systems. *Nature* 410:926-930.

Reganold, J. P., R. I. Papendick, and J. F. Parr (1990). Sustainable agriculture. *Scientific American* 261:112-120.

U.S. Department of Agriculture, Agriculture Marketing Service (2001). *National organic policy,* 7 CFR part 205. Washington, DC: USDA.

1. TEMPERATURE RELATIONS
2. TRAINING AND PRUNING
PRINCIPLES
3. TRAINING SYSTEMS
4. TREE CANOPY TEMPERATURE
MANAGEMENT

– 1 –

Temperature Relations

Rajeev Arora

Low winter temperatures are a major limiting factor in the production of temperate tree fruit crops. Economic losses can occur as a direct result of severe midwinter temperatures or untimely fall or spring frosts. For these reasons, cold hardiness is often a selection criterion in breeding programs, and several research programs have placed a primary emphasis on elucidating the mechanisms of freezing injury and winter survival of fruit crops. This chapter attempts to provide a brief overview of the subject and to discuss various strategies that are currently being used to study freezing injury and winter survival.

FREEZING TOLERANCE VERSUS FREEZING AVOIDANCE

Strategies that allow plants to survive freezing temperatures have been placed into two major categories: freezing tolerance and freezing avoidance. Tissues displaying *freezing tolerance* respond to a low-temperature stress by the loss of cellular water to extracellular ice. This results in collapse of the cell wall (cytorrhysis) and increased concentration of the cell sap, which, in turn, lowers the freezing point. In contrast, tissues that *avoid* freezing stress but are still exposed to freezing temperatures do so by deep supercooling, a process in which cellular water is isolated from the dehydrative and nucleating effects of ice present outside the cell in extracellular spaces. It is noteworthy that, in many woody perennials, different tissues (bark and leaves versus xylem and buds) within the same plant respond in distinctly different ways to freezing temperatures that represent both aforementioned strategies.

Deep Supercooling of Xylem Tissues

Deep supercooling occurs in more than 240 species in 33 families of angiosperms and one family of gymnosperms (Quamme, 1991). Many deciduous tree fruit crops, such as apple, apricot, cherry, peach, pear, and plum, are known to exhibit deep supercooling. Living xylem tissues in most of these fruit crops avoid low temperature stress by deep supercooling. During the supercooling event, cellular water remains liquid within xylem parenchyma cells at very low temperatures by remaining isolated from heterogeneous ice nuclei and the nucleating effect of extracellular ice. Supercooled water is in a metastable condition and will form intracellular ice in response to a heterogeneous nucleation event or when the homogeneous nucleation temperature of water ($-38°C$) is reached. If and when intracellular freezing occurs, it always results in death of the tissue.

The freezing response of xylem tissues of fruit trees can be monitored using the technique of differential thermal analysis (DTA). Thermocouples are utilized to detect the heat of fusion produced by water in the samples as it undergoes a liquid to solid phase change. In DTA, sample temperatures are compared to a piece of freeze-dried tissue (reference) undergoing the same rate of cooling. This produces a flat baseline until the freezing of water within the sample tissue results in a difference in temperature between the sample and the reference. The sample-reference differential is visualized as a peak on a thermogram, hence the term "differential thermal analysis."

In thermograms of woody plants that exhibit deep supercooloing (Figure T1.1), the initial large peak is referred to as a high-temperature exotherm (HTE) and is believed to represent the freezing of bulk water contained within tracheary elements and extracellular spaces, whereas the peak occurring at very low temperatures (deep supercooling) is believed to represent the freezing of intracellular water contained within xylem parenchyma cells. The peak resulting from the freezing of deep supercooled water is referred to as a low-temperature exotherm (LTE). Typically, LTE on these thermograms is correlated with the death of xylem ray parenchyma cells. Because of this association with mortality, DTA has been extensively used to evaluate the degree of cold hardiness of stem tissues of fruit trees. DTA is also used to detect seasonal changes in stem hardiness since woody plants display distinct seasonality of supercooling ability, in that it in-

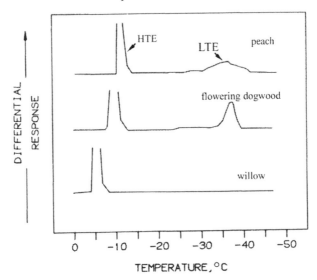

FIGURE T1.1. Freezing response of internodal xylem (debarked twig) of peach, flowering dogwood, and willow subjected to differential thermal analysis. HTE: high-temperature exotherm; LTE: low-temperature exotherm. Willow, a non-supercooling species, lacks LTE.

creases and decreases in fall and spring, respectively, and is greatest in winter. Many woody plants, including some fruit crops, do not exhibit deep supercooling. In these species, a typical DTA thermogram lacks LTE (Figure T1.1).

For deep supercooling to occur, a tissue must exhibit several features: (1) cells must be free of heterogeneous nucleating substances that are "active" at warm subfreezing temperatures; (2) a barrier must be present that excludes the growth of ice crystals into a cell; and, concomitantly, (3) a barrier to water movement must exist that prevents a "rapid" loss of cellular water to extracellular ice in the presence of a strong vapor pressure gradient. It is believed that physical properties of the cell wall largely account for the ability of xylem parenchyma cells to deep supercool (Wisniewski, 1995). In this regard, the use of colloidal gold particles of prescribed sizes and other apoplastic tracers have been used to study the porosity and permeability of cell walls. Results of these studies indicate that the pit membrane (a thin portion of the cell wall that allows for the passage of solutes, as well as plasmodesmatal connections, between cells) and the associated amorphous layer, not the secondary wall, may play a limiting role in determining the ability of the cell wall to retain water against a strong vapor pressure gradient and the intrusive growth of ice crystals. Results from studies with peach xylem tissues indicate that by chemically or enzymatically altering the structure of the pit membrane, which is mainly composed of cellulose and pectic materials, the extent of deep supercooling can be reduced or eliminated (Wisniewski, 1995).

Although pectin-mediated regulation of deep supercooling may account for the aforementioned observations, many fundamental questions must still be resolved. For example, how do species that supercool differ from those that do not? How do we account for seasonal shifts in deep supercooling? If pectin degradation (or lack thereof) is a key determinant of seasonal changes in supercooling ability, do changes in the activity and/or amount of pectin-degrading enzymes parallel these seasonal shifts in supercooling ability? Furthermore, it has been reported that apple, peach, and some other species do not exhibit homogeneous freezing responses, in that their thermograms show multiple LTEs (in apple) or bimodal peaks (in peach). How are these complex freezing behaviors regulated? In-depth investigations aimed at answering these questions will indicate whether this trait can be manipulated in a manner that will enhance cold hardiness in tree fruit crops.

Extraorgan Freezing and Deep Supercooling of Dormant Buds

The response of dormant buds of tree fruit to freezing temperatures is different from that of other portions of the tree and varies with species. The pattern of freezing in the dormant floral buds of apple and pear begins with the initiation of freezing of extracellular water within the bud scales and subtending stem tissues. The introduction of ice into the bud tissue results in the establishment of a water potential gradient. Consequently, water migrates from the shoot or floral apex to the sites of extracellular ice in response to the water potential gradient. Thus, the floral primordium is isolated from the mechanical damage caused by the presence of large ice crystals. This response to freezing temperatures has been described as "extraorgan freezing" (Sakai, 1979). When buds are killed, mortality results from the dehydrative stress rather than the low temperature or the presence of ice.

Extraorgan freezing is characteristic of most cold-hardy species, and vegetative buds of all temperate fruit species respond in this manner.

In other fruit species, however, not all the water from floral tissue migrates to the ice in bud scales. Instead, a portion of water remains supercooled within the floral tissue. Deep supercooling of flower buds has been observed in several *Prunus* species (Quamme, 1991). For these species, as with deep supercooling of xylem tissues, two distinct exotherms are detected when buds are subjected to DTA. The HTE is associated with the freezing of water in the bud scales and subtending stem tissue, and the LTE is associated with the freezing of intracellular, deep supercooled water contained within the floral tissue. The LTE is correlated with the degree of cold hardiness of tissue and is used extensively as an evaluation tool. In species containing multiple flowers within a single flower bud, each floret freezes as an independent unit. This is demonstrated by the appearance of multiple LTEs obtained using DTA. This is true for sweet cherry and sour cherry (Quamme, 1995).

As in the case of xylem parenchyma cells, in order for deep supercooling of buds to occur, a barrier to water movement and ice propagation must exist. The nature of this barrier and how deep supercooling in buds is regulated are not fully understood. However, research with peach flower buds suggests that the loss of deep supercooling (during the spring when the buds begin to break dormancy and progressively lose cold hardiness) is associated with the development of vascular continuity between the flower and stem axis (Ashworth, 1984). The functional strand of xylem between the developing bud primordium and the subtending shoot serves as a conduit for the rapid spread of ice into the primordium, and deep supercooling can no longer occur. Whether similar events occur in other fruit species is not known.

Equilibrium Freezing

In contrast to xylem tissue of some species and some dormant buds, which avoid freezing stress by supercooling, bark and leaf tissues of temperate fruit trees undergo "equilibrium freezing" and concomitantly tolerate the extracellular ice formation and the dehydrative stress that result from the loss of cellular water to extracellular ice. Equilibrium freezing in plant cells occurs during slow cooling rates (1 to 2°C per hour), when ice formation is initiated at high subzero temperatures (−1 to −4°C) in the extracellular spaces due to a nucleation event. This occurs because (1) the extracellular solution has a higher (warmer) freezing point than the intracellular solution or cell sap and (2) efficient nucleators such as dust or bacteria are prevalent in the extracellular environment. Once the tissue temperature drops below the freezing point of the cell sap, the internal vapor pressure becomes higher than that of extracellular ice. The formation of this gradient results in the movement of cellular water to extracellular ice crystals, which then increase in size. A gradual or slow cooling allows the diffusion of cellular water to ice at a speed sufficient to increase the solute concentration of the cell sap as rapidly as the temperature drops. This allows the chemical potential of the cell sap to be in equilibrium with that of the ice, hence, the term "equilibrium freezing." As a result of this type of freezing, the leaf and bark cells of fruit trees undergo dehydrative stress (due to the loss of cellular water), low temperature stress per se, mechanical stress (due to the presence of large ice crystals and cell wall collapse), and toxic stress (due to the increased concentration of solutes in the cell sap).

OVERWINTERING

Winter survival of woody perennials, including temperate tree fruit crops, is dependent on two phenological events: (1) the onset of dormancy during fall and (2) an ability to increase freeze tolerance upon exposure to low nonfreezing temperatures (e.g., a change from a freeze-susceptible to a freeze-resistant state—a process called cold acclimation). Once plants are in a dormant state, an exposure to a chilling period is required for floral and vegetative budbreak in the following spring. Chilling requirement prevents growth from occurring during periodic warm spells during winter and thus helps synchronize plant growth with the prevalence of favorable environmental conditions. Cold acclimation (CA), on the other hand, enables plants to survive the subfreezing temperatures present during winter. Due to the process of cold acclimation, woody plant tissues that would be killed by temperatures slightly below 0°C during summer and early fall may survive temperatures as low as −70°C during winter. The extent to which a particular species can acclimate is largely genetically determined. In the spring during an annual cycle, cold-acclimated plants begin to lose their acquired hardiness (deacclimation) while they also come out of the dormant state. Hence, transitions (onset and loss) in dormancy and cold hardiness partially overlap.

In nature, cold acclimation in woody plants is typically a two-stage process. The first stage (initial increase in freezing tolerance) is induced by short days, whereas the second stage is induced, primarily, by low temperatures. Therefore, full cold acclimation potential (maximum midwinter freezing tolerance) is normally achieved as an additive response of tree fruit crops to both environmental cues. Dormancy or rest in many woody perennials is also induced or enhanced by short photoperiods in the fall. A deviation from the aforementioned environmental regulation of cold acclimation and/or dormancy in woody plants, albeit possible, is considered an atypical response, and those genotypes which exhibit such differences often serve as valuable experimental systems to gain fundamental knowledge of the environmental physiology of these two processes.

The seasonal shifts in dormancy and cold hardiness status during the annual cycle of tree fruit crops imply that these are active processes of adaptation. Research indicates that the two processes are genetically regulated and lead to changes in metabolism and cell composition. Among the transformations that occur during overwintering of woody plants are distinct shifts in gene activity, changes in carbohydrate metabolism that lead to accumulation of specific types of sugars, changes in the composition of cell membranes, accumulation of abscisic acid (ABA, a plant growth hormone), and accumulation of unique classes of proteins (Chen, Burke, and Gusta, 1995). However, the simultaneous occurrences of dormancy and cold hardiness transitions make it difficult to associate physiological and molecular changes that specifically control one or the other phenological event. To overcome this problem, researchers have devised several physiological and/or genetic approaches and strategies (Wisniewski and Arora, 2000) that are briefly described here.

SYSTEMS AND STRATEGIES TO DISTINGUISH COLD HARDINESS AND DORMANCY TRANSITIONS

Use of Sibling Deciduous and Evergreen Genotypes

One of the first attempts to study protein changes associated specifically with the changes in cold hardiness or dormancy was through the use of sibling genotypes of peach *(Prunus persica),* segregating for deciduous and evergreen habits. The deciduous genotype typically enters dormancy during fall and exhibits CA. The evergreen genotype, on the other hand, exhibits CA, but the apical meristem of these trees remains nondormant throughout the seasonal cycle. Researchers have characterized seasonal patterns of cold hardiness and protein profiles in bark tissues of these genotypes. Comparative analyses of the seasonality and the degree of CA with that of protein changes in the two genotypes (one lacking dormancy, whereas the other not) have enabled these researchers to specifically associate certain protein changes with CA and others with dormancy (Figure T1.2).

Differential Induction of Dormancy and Cold Acclimation

Researchers have also used systems in which the developmental program of dormancy can be induced separately from CA. For exam-

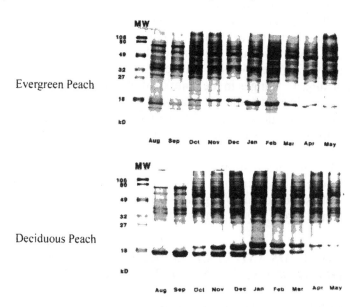

FIGURE T1.2. Monthly profiles of bark proteins (separated by gel electrophoresis) of sibling deciduous and evergreen peach trees. Note qualitative and quantitative protein differences between the seasonal patterns for the two genotypes. (*Source:* Modified from Arora, Wisniewski, and Scorza, 1992.)

ple, *Vitis labruscana,* a grape species, exhibits a rather unique developmental programming, in that it is able to fully enter dormancy in response to short photoperiods without cold acclimating. By employing controlled-environment treatments, researchers have exploited this system to characterize differential accumulation of proteins in grape buds during superimposed dormancy and CA programs (use of short photoperiods and cold treatment), and in the buds that had entered only the dormancy program (use of only short photoperiods). By analyzing the profiles of bud proteins from these treatments, they have identified gene products (proteins) that are specific to cold acclimation and those specific to dormancy development.

Differential Regulation of Chill Unit Accumulation and Cold Hardiness

Chilling requirement (CR), a genetically determined trait, is defined as the need for exposure to low temperatures for a genetically determined period of time in order for buds to overcome dormancy and resume normal growth the following spring. The CR of a species is described as the number of hours (chill units, or CUs) of low-temperature exposure needed, and the progress toward meeting the requirement, as the chill unit accumulation (CUA). Temperatures of 0 to 7°C, which also induce cold acclimation, are typically considered to contribute toward CUA. However, temperatures above and below that range do not contribute to CUA (Rowland and Arora, 1997). Exposure to relatively warmer temperatures (10 to 15°C) may also cause cold-acclimated buds to deacclimate in certain species without negating CUA (dormancy neutral treatment). This premise has been used by researchers as the basis to differentially modify CUA and cold hardiness transitions in certain fruit crops and, thereby, to identify physiological changes specifically associated with these events.

Dehydrins

Biochemical and molecular studies of cold acclimation in plants have led to the discovery that numerous environmental cues (dehydration, low temperature, increased concentration of cell sap) and treatment with ABA induce accumulation of a similar class of proteins called "dehydrins" (Close, 1997). A functional role for dehydrins in freezing tolerance of plants is suggested, in part, by their hydrophilic properties (thereby protecting cell membranes and other organelles from desiccation), and follows the logic that, since plant cells undergo dehydration during freezing stress, the cellular responses invoking desiccation tolerance should also be involved in freezing tolerance mechanisms. Biochemical and physiological studies employing the aforementioned three strategies show the accumulation of specific dehydrins in cold-acclimated tissues of certain fruit crops (Wisniewski and Arora, 2000, and references therein). Moreover, data also indicate that these dehydrins typically accumulate at much higher levels in hardier siblings, cultivars, and tissues compared to less hardy ones (Rowland and Arora, 1997). However, studies to date have only established correlative relationships between dehydrin accumulation and increased cold hardiness, and no "cause and effect" relationship has yet been established in tree fruit crops. Whereas there is little doubt that dehydrins are key biochemical factors in the cold acclimation process, their specific role in increasing a plant's freezing tolerance remains to be unraveled and will likely be a subject of future investigations.

Responses of fruit trees to low temperatures are both varied and complex. Different tissues within the same plant respond differently to subzero temperatures. This is further complicated by the fact that fruit trees undergo seasonal transitions in cold hardiness that are superimposed by changes in dormancy status of the plant. The economic importance of fruit production and limits placed on it due to low temperature stress, however, will ensure that research continues to develop better understanding of the adaptation and response of fruit trees to cold temperatures.

Related Topics: DORMANCY AND ACCLIMATION; FLOWER BUD FORMATION, POLLINATION, AND FRUIT SET; GEOGRAPHIC CONSIDERATIONS; SPRING FROST CONTROL

SELECTED BIBLIOGRAPHY

Arora, R., M. W. Wisniewski, and R. Scorza (1992). Cold acclimation in genetically related (sibling) deciduous and evergreen peach (*Prunus persica* L.

Batsch). I. Seasonal changes in cold hardiness and polypeptides of bark and xylem tissues. *Plant Physiol.* 99:1562-1568.

Ashworth, E. N. (1984). Xylem development in *Prunus* flower buds and its relationship to deep supercooling. *Plant Physiol.* 74:862-865.

Chen, T. H. H., M. J. Burke, and L.V. Gusta (1995). Freezing tolerance in plants. In Lee, R. E., C. J. Warren, and L. V. Gusta (eds.), *Biological ice nucleation and its applications* (pp. 115-136). St. Paul, MN: APS Press.

Close, T. J. (1997). Dehydrins: A commonality in the response of plants to dehydration and low temperatures. *Plant Physiol.* 100:795-803.

Quamme, H. A. (1991). Application of thermal analysis to breeding fruit crops for increased cold hardiness. *HortScience* 26:513-517.

Quamme, H. A. (1995). Deep supercooling of buds in woody plants. In Lee, R. E., C. J. Warren, and L. V. Gusta (eds.), *Biological ice nucleation and its applications* (pp. 183-200). St. Paul, MN: APS Press.

Rowland, L. J. and R. Arora (1997). Proteins related to endodormancy (rest) in woody perennials. *Plant Science* 126:119-144.

Sakai, A. (1979). Freezing avoidance mechanism of primordial shoots of conifer buds. *Plant Cell Physiol.* 20:1381-1386.

Wisniewski, M. (1995). Deep supercooling in woody plants and role of cell wall structure. In Lee, R. E., C. J. Warren, and L. V. Gusta (eds.), *Biological ice nucleation and its applications* (pp. 163-181). St. Paul, MN: APS Press.

Wisniewski, M. and R. Arora (2000). Structural and biochemical aspects of cold hardiness in woody plants. In Jain, S. M. and S. C. Minocha (eds.), *Molecular biology of woody plants* (pp. 419-437). Dordrecht, the Netherlands: Kluwer Academic Publishers.

SHOOT ORIENTATION EFFECT ON APICAL DOMINANCE

Apical dominance is strongest in the actively growing terminal buds of vertical shoots and limbs. Thus, limb orientation has a dramatic effect on apical dominance and, thereby, the pattern of vegetative and reproductive growth. A vertical shoot would tend to be the most vegetative. As limb orientation shifts toward horizontal, terminal shoot growth decreases while number and length of lateral shoots increase. Generally, when limbs or shoots are oriented at 30 to 60 degrees from vertical, vegetative shoot growth in the terminal area is reduced while the number and length of lateral shoots farther away from the terminal are increased. As such, more moderate limb or shoot orientations have the potential to provide a balance between terminal and lateral shoot development as well as promote the development of flower buds. When limbs or shoots become at or below horizontal, however, the influence of apical dominance is lost. As a result, the lateral shoots that are normally influenced by apical dominance can develop unchecked into vigorous, upright water sprouts. These water sprouts then become independent vertical shoots with strong apical dominance.

Balancing vegetative growth and flower bud production is the basis for the limb spreading or positioning that is commonly practiced in tree fruit production. Response varies with cultivar, rootstock, tree age, degree of spreading, and time of spreading. If both cultivar and rootstock are precocious (prone to early and heavy fruiting), extreme limb spreading may result in excessive flower bud production and insufficient shoot growth, whereas, with more vigorous cultivars, wide limb spreading can result in severe loss of flowers due to production of water sprouts. Time of spreading or positioning also affects response. Spreading in late season after growth has terminated will have little effect on vegetative growth during that season. Conversely, extreme spreading in the dormant season before growth begins can result in weak terminal shoot growth and excessive water sprout development. Cultivars vary in response to limb orientations. For example, spur-type 'Delicious' trees are prone to water sprout development when scaffolds are oriented more than 60 degrees from vertical, whereas well-branched cultivars such as 'Golden Delicious' are not.

GENERAL RESPONSES TO PRUNING

As pruning removes vegetative material, it changes the balance between the aboveground part of the tree (shoots, buds, and leaves) and the belowground part (roots). When pruning alters this balance, the tree responds with vegetative regrowth until the balance is reestablished. This stimulation of regrowth is in close proximity to the cuts. The amount of regrowth that follows pruning is in direct proportion to the severity of pruning (Table T2.1), with vegetative growth developing at the expense of flower bud formation. As pruning severity increases, flower bud production decreases proportionally (Figure T2.1). Pruning young trees delays the onset and amount of early fruit production. Excessive pruning in bearing trees can result in excessive vegetative vigor, a reduction in spur and flower development, and a decrease in yield. As shoot growth increases following pruning, root growth is decreased. The reestablishment of the balance within the tree results from an increase in shoot development and a slowdown in root development. Ultimately, pruning reduces the total growth of the tree and, as such, remains a major method of tree size control in fruit production.

TYPES OF PRUNING CUTS

The two basic types of pruning cuts are heading and thinning. Each results in different growth responses and has specific uses.

TABLE T2.1. Influence of pruning severity on growth of 'Delicious' apple trees

		Shoot growth/limb	
Pruning severity	Shoot number	Average length (cm)	Total growth (cm)
0	20.4	19.6	402
1	16.0	23.6	361
2	14.9	24.8	362
3	8.4	29.6	244

Source: Modified from Barden, DelValle, and Myers, 1989.

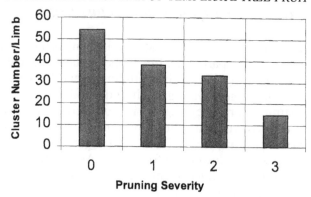

FIGURE T2.1. Influence of pruning severity on flower clusters of 'Delicious' apple (*Source:* Modified from Barden, DelValle, and Myers, 1989.)

Heading removes the terminal portion of a shoot or limb. By removing apical dominance, heading stimulates growth near the cut. Heading is the most invigorating type of pruning cut, and the shoot or shoots that develop immediately below the cut reestablish apical dominance. Heading cuts are the most disruptive to the natural growth and form of trees, although they are useful to induce branching at specific points, such as in establishing scaffolds. In production systems where early fruit production is critical for economic return, use of heading cuts should be kept to a minimum.

Thinning, on the other hand, removes an entire shoot or limb to its point of origin from a main branch or limb. Thinning may also include the removal of a shoot back to a lateral shoot or spur. With thinning cuts, some terminal shoots are left intact, apical dominance remains, and the pruning stimulation is more evenly distributed among remaining shoots. New growth is dominated by the undisturbed shoot tips, while lateral bud development follows more natural patterns for that species or cultivar. Thinning cuts are generally the least invigorating and provide a more natural growth form for trees. Important in maintenance pruning, thinning cuts are used to shorten limbs, improve light penetration into tree canopies, and direct the growth of shoots or limbs. Studies show that heading cuts result in high numbers of shoots and reductions in fruit, whereas thinning cuts increase fruit number and control vegetative growth.

TRAINING AND PRUNING OBJECTIVES

Young, Nonbearing Trees

In the young, nonbearing tree, the focus of management is development of tree structure with the objective of filling the allotted canopy space within the given orchard system. Light pruning is more desirable than heavy pruning during this period, as heavily pruned trees exhibit less increase in trunk and root growth than trees that are lightly pruned. Pinching the tips of developing laterals in apple results in a decrease in total shoot growth as severity of pruning increases. Although remaining shoots are significantly longer, shoot number is decreased.

Positioning of limbs influences the subsequent development of vegetative growth and lays the groundwork for the development of fruiting wood. Vertical limbs develop relatively few laterals, with number of laterals increasing as orientation moves from vertical to horizontal. Although more shoots develop on horizontal limbs, average shoot length is reduced. Positioning limbs at moderate angles allows an increased number of laterals to develop while minimizing reduction in shoot length.

Young, Bearing Trees

During this period, management of tree resources expands to include initial development of fruiting sites and the initial phase of flowering, as well as continued expansion of tree structure to fill allotted space. Heavy pruning can delay the filling of the allotted canopy area within the orchard by reducing total shoot number and total shoot growth. Pruning severity can also delay flowering, as flower cluster number decreases as pruning severity increases. Research indicates that pruning should be minimized in order to maximize the potential for early flowering and fruiting. In order to encourage branching, limbs may be positioned at desirable angles.

In addition, bagging of the central leader of apple is useful in some cultivars to encourage a higher number of shoots to develop. The practice also increases the number of flower clusters and subsequent number of fruit. Notching immediately above buds is another technique for increasing lateral development. Notching is equally effective on all sections of the limb, with over 90 percent of notched buds developing into shoots. This effect has the highest impact in the basal and middle sections of the limb where natural lateral branching occurs less frequently. The production of fruit has a tremendous impact on shoot growth and future fruit production. There is a limited supply of resources within the tree for growth. With the presence of fruit, there is a decrease in the supply of resources available for shoot and root growth. As a result, shoot and root growth will decrease relative to the level of fruit production.

Mature, Bearing Trees

After a tree's desired canopy size has been attained, management begins to focus on the maintenance of fruiting sites to provide sustained flower and fruit production along with the annual expansion of associated shoot growth. The focus is on merging the appropriate pruning and training techniques necessary to maintain tree size while keeping vegetative growth in balance with the need to maximize fruit quality and yield efficiency. Appropriate pruning cuts are used primarily to remove weak, unproductive wood and enhance light penetration into the canopy; ideally, pruning and training techniques are utilized to maximize continual development of fruiting wood throughout the tree canopy. Maintenance and renewal of fruiting sites are critical to sustained fruit quality as a tree ages. Additional pruning focuses on the prevention or removal of excessive shoot growth that can develop in localized areas and reduce light penetration into the canopy. For example, removal of strong, upright water sprouts within the canopy of mature peach trees approximately four weeks prior to harvest increases fruit size and flower bud formation (Figure T2.2).

Growth and development in tree fruit are regulated by certain basic biological processes. Training and pruning practices have a significant impact on the balance between vegetative growth and reproductive growth. An understanding of the biological principles that govern growth and development as well as the basic principles underlying the effects of pruning and training will permit the orchardist to balance vegetative and reproductive growth in the most efficient, effective, and practical manner.

Related Topics: CARBOHYDRATE PARTITIONING AND PLANT GROWTH; CULTIVAR SELECTION; DWARFING; FLOWER BUD FORMATION, POLLI-

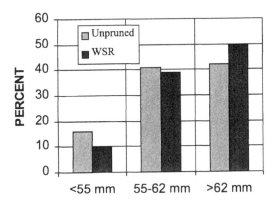

FIGURE T2.2. Effect of preharvest water sprout removal (WSR) on packout of 'Redskin' peach (*Source:* Modified from Myers, 1993.)

NATION, AND FRUIT SET; LIGHT INTERCEPTION AND PHOTOSYNTHE-SIS; PLANT GROWTH REGULATION; PLANT HORMONES; ROOTSTOCK SELECTION; TRAINING SYSTEMS

SELECTED BIBLIOGRAPHY

Barden, J. A., T. B. G. DelValle, and S. C. Myers (1989). Growth and fruiting of Delicious apple trees as affected by severity and season of pruning. *J. Amer. Soc. Hort. Sci.* 114:184-186.

Forshey, C. G., D. C. Elfving, and R. L. Stebbins (1992). *Training and pruning apple and pear trees.* Alexandria, VA: Amer. Soc. Hort. Sci.

Maib, K. M., P. K. Andrews, G. Lang, K. Mullinix, eds. (1996). *Tree fruit physiology: Growth and development.* Yakima, WA: Good Fruit Grower, Wash. State Fruit Commission.

Myers, S. C. (1988). Basics in open center peach tree training. In Childers, N. F. and W. B. Sherman (eds.), *The Peach* (pp. 389-403). Somerset, NJ: Somerset Press, Inc.

Myers, S. C. (1993). Preharvest watersprout removal influences canopy light relations, fruit quality, and flower bud formation of Redskin peach trees. *J. Amer. Soc. Hort. Sci.* 118:442-445.

Rom, C. R. (1989). Physiological aspects of pruning and training. In Peterson, A. B. (ed.), *Intensive Orcharding* (pp. 13-40). Yakima, WA: Good Fruit Grower.

Tustin, S. (1991). Basic physiology of tree training and pruning. *Proc. Wash. State Hort. Assoc.* 87:50-63.

3

Training Systems

Tara Auxt Baugher

Tree architecture impacts photosynthetic potential and partitioning efficiency. Many training systems have been developed, and each has a place, depending on rootstock, cultivar, environmental conditions, and grower preferences. For example, the slender spindle system is best used with a full dwarf rootstock and was originally adopted in European regions with low light conditions. Training systems offer specific opportunities, and growers can choose from designs that increase productivity, improve fruit color and quality, augment integrated production efforts, boost work efficiency, and/or complement market strategies.

CENTRAL LEADER SYSTEMS

Central Leader

Central leader–trained trees are adaptable to a wide range of environmental and socioeconomic conditions. The general tree shape is conical, with the largest diameter branches at the bottom and the smallest at the top. D. R. Heinicke popularized the system in the United States, with a "head and spread" training concept. This system of heading the central leader and scaffolds to encourage branching is particularly effective on spur-type 'Delicious' apple trees. D. W. McKenzie developed a similar system in New Zealand, a difference being that bays are created for ladder placement and ease of harvesting. With standard or semidwarf rootstocks, trees are freestanding, and there are three tiers of scaffolds on the leader with gaps in between for sunlight penetration. In recent years, growers have modified the system for use with staked dwarf trees. Trees grow together in some systems and are managed as hedgerows. Individual identity is lost, but from the ends of the rows trees should appear conical. Although pyramid training has been most widely tested on apples, central leader and modified central leader systems also are used on peaches, nectarines, cherries, plums, and pears. Due to growth habit, some stone fruit cultivars are better adapted to a central leader system than others. A disadvantage with pear trees is the potential for losing the leader to a fire blight strike. A successful alternative is to train trees to a four-leader system.

Slender Spindle

Various forms of the slender spindle system are highly productive in Europe and other fruit-growing areas. The system was developed in Holland and Belgium in the 1950s and described by S. J. Wertheim in 1970 in a Wilhelminadorp Research Station publication. It was quickly adopted in Germany and other countries with limited land for agriculture. Tree height is 2.0 to 2.5 meters, and only the bottom whorl (or table) of branches is permanent. Higher, fruiting branches are kept weak by bending and are regularly renewed by strategic pruning. Trees are individually staked, and branches are tied down to induce fruiting and tied up to support crop loads. Slender spindle trees are 1.0 to 1.5 meters at the base and taper to a point at the top. Super slender spindle trees are even narrower. Slender spindle–trained trees can be grown in multiple-row (bed) systems; however, most growers prefer single rows. Economic success of a slender spindle orchard depends on proper selection of a precocious, size-restricting rootstock, well-feathered nursery stock, and detailed training and pruning. Slender spindle–type systems for stone fruit trees are under test in some regions. The fusetto is an Italian hedgerow system that utilizes slender spindle–trained peach trees.

Vertical Axis

The vertical axis is the system of choice for many apple growers who are making a transition from low-density, central leader–trained trees to higher-density production systems. J.-M. Lespinasse developed the system in the 1970s and described it in a French apple tree management bulletin published in 1980. The system is designed to encourage equilibrium between fruit production and vegetative growth. Trees receive minimal pruning and training, which results in early production and natural growth control. A training pole attached to one or two trellis wires supports the leader of each tree. The bottom whorl of limbs is permanent, and upper branches are periodically renewed. Widely used rootstocks are M.9 and M.26, and tree height ranges from 3.5 to 5.0 meters, depending on stion (scion/rootstock) and site. Although ladders must be used for picking, harvest is more efficient due to narrow tree widths. A recent modification of the vertical axis is the solaxe, in which renewal pruning is replaced by bending.

Slender Pyramid

A number of New Zealand growers train apple trees to a slender pyramid, which is a hybrid of the central leader and the vertical axis system. D. S. Tustin is one of the originators. Rapid canopy development and early fruiting are encouraged by avoiding dormant heading cuts and by strategic summer pinching. Basic tree form is pyramidal with a strong basal tier of scaffolds and a slender upper canopy. Commonly used rootstocks are M.26 and MM.106. Trees are only slightly narrower than central leader–trained trees, but due to New Zealand's ideal growing environment, production is equal to that from more intensive plantings in other regions.

Hybrid Tree Cone

The hybrid tree cone (HYTEC) combines some of the best characteristics of the vertical axis and the slender spindle. In the 1990s, B. H. Barritt developed the hybrid conical tree form for central Washington State cultivars and climatic conditions. Various modifications of the HYTEC have been successfully applied in other fruit-growing regions. Trees are staked and are intermediate in height between the slender spindle and the vertical axis. To reduce excessive vigor in the top of the tree, the central leader is zig-zagged by annual bending or removal to a side branch. Otherwise, training is similar to that used with the vertical axis system.

OPEN CENTER SYSTEMS

Open Vase

Most stone fruit cultivars have a spreading growth habit and are easily trained to a freestanding open vase. The system is widely planted in many stone fruit production areas because it is easy to manage and sunlight is intercepted throughout the day. Three to four wide-angled, evenly spaced scaffolds are selected off a short trunk. The primary scaffolds are allowed to branch into secondary and tertiary branches, with fruiting wood developed in all sections of the canopy. General tree shape varies depending on microclimate and cultivar characteristics. For instance, California peach trees are trained to taller, more upright forms than eastern U.S. trees; some plum cultivars, e.g., 'Santa Rosa', are more spreading than others, e.g., 'Wickson'. Although open center training is currently the most widely used system with stone fruit, this may change as breeders introduce new size-restricting rootstocks and tree forms. Several growth types developed through U.S. Department of Agriculture breeding programs are the spur-type, weeping, and columnar forms.

Delayed Vase

An increasing number of peach trees in Italy are trained to a delayed vase. Trees are handled in a manner similar to the open vase system, except that a weak leader is maintained during the initial years of training. The presence of shoots in the center of the tree forces the scaffolds to develop stronger, wider angles. The leader is progressively weakened and can be removed at the end of the third growing season. In the United States, S. C. Myers developed a similar concept of training known as "topped center." Two to four dominant shoots in the tree center are maintained but continually weakened through summer tipping. The training strategy helps growers avoid a common problem with either vase or V systems, which is the undesirable use of bench cuts to correct for narrow-angled crotches.

Supported V

Training trials on pome and stone fruit demonstrate that V-shaped systems offer the potential to intercept more sunlight than vertical systems. One of the first V trellises to be widely tested was the Tatura. D. J. Chalmers and B. van den Ende conducted the original studies at the Tatura Research Station in Australia. Trellis frame angle is 60 degrees. Individual trees are trained to a V, or alternating trees are leaned to one side or the other along the row. Research shows that a thin canopy must be maintained for optimum light distribution. A modification of the Tatura is the MIA (for Murrumbidgee Irrigation Area), developed by R. J. Hutton. The trellis is an inverted Tatura that permits access on both sides of the canopy. Other successful supported V systems include the Y trellis tested extensively in New York, the Güttinger V developed in Switzerland, and V and Y spindles, in which each tree is trained to a spindlebush.

Freestanding V

Freestanding V systems offer many of the advantages of V trellises but at reduced costs. Stone fruit growers in several regions are adopting various perpendicular V systems in efforts to increase production and management efficiency. Two that originated in California are the Kearney V and the Quad V. Research by T. M. DeJong and K. R. Day demonstrates that standard-size trees can be more intensively managed when trained to perpendicular V forms. The Kearney V is a Tatura without a trellis, except that windows for lateral light penetration are maintained between trees. The Quad V is essentially a "double" Kearney V. Instead of two scaffolds growing out into the row middles, there are four.

VERTICAL AND HORIZONTAL TRELLIS SYSTEMS

Vertical Canopies

The palmette trellis originated in Italy and France and is most commonly used in areas where growers prune and harvest their trees from moveable platforms. Trellis height is 3.0 to 4.5 meters. Trees are trained to a two-dimensional form, with branches tied to a horizontal or an oblique position. An adaptation for dwarf apple trees is the Penn State low trellis hedgerow, developed by L. D. Tukey. Several interesting vertical trellises—marchand, drapeau, and Belgian cordon—

orient trunks and limbs at 45 degrees to balance cropping and shoot growth.

Horizontal Canopies

The Lincoln canopy is a perpendicular T-shaped trellis. J. S. Dunn, a New Zealand agricultural engineer, designed the system for mechanical harvesting. Detailed training and pruning are recommended to prevent low-light conditions. Early tests with containment spraying were conducted on this system. The solen is a system developed by J.-M. Lespinasse for acrotonic apple growth habits. The system resembles a T from the side, rather than from the end of the row. Main scaffolds are trained horizontally along two wires, and fruit hang underneath. A horizontal, suspended canopy system, or pergola, is used in Japan for pears.

A considerable body of knowledge is available on the performance of tree fruit training systems. Many of the studies show that when rootstock and tree spacing are constant and level of pruning is minimal, early productivity is similar. Degree of success often is due to how well a system is tailored to specific climatic conditions and management objectives. As cultivars change, flexibility also becomes important. A current thrust in training research is to assess natural tree growth habits and fruit quality requirements in order to help growers find the best match between a cultivar and a training system. The systems (Figure T3.1) described in this section are the more common ones found in pomology literature. Many innovative orchardists modify training strategies or develop new designs to maximize system potential.

Related Topics: CULTIVAR SELECTION; DWARFING; GEOGRAPHIC CONSIDERATIONS; HARVEST; HIGH-DENSITY ORCHARDS; LIGHT INTERCEPTION AND PHOTOSYNTHESIS; ROOTSTOCK SELECTION; TRAINING AND PRUNING PRINCIPLES

SELECTED BIBLIOGRAPHY

Barritt, B. H. and F. Kappel, eds. (1997). *Proceedings of the sixth international symposium on integrating canopy, rootstocks and environmental physiology in orchard systems,* No. 451. Leuven, Belgium: Acta Horticulturae.

Baugher, T. A., S. Singha, D. W. Leach, and S. P. Walter (1994). Growth, productivity, spur quality, light transmission and net photosynthesis of 'Golden Delicious' apple trees on four rootstocks in three training systems. *Fruit Var. J.* 48:251-255.

Day, K. R. and T. M. DeJong (1999). Orchard systems for nectarines, peaches and plums: Tree training, density and rootstocks. *Compact Fruit Tree* 32:44-48.

Ferree, D. C. (1994). Orchard Management Systems. In Arntzen, C. J. and E. M. Ritter (eds.), *Encyclopedia of agricultural science* (pp. 131-142). San Diego, CA: Academic Press, Inc.

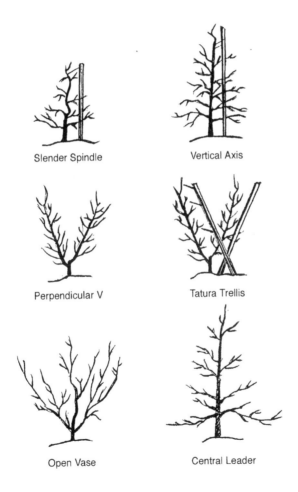

FIGURE T3.1. Fruit tree training systems

4

Tree Canopy Temperature Management

D. Michael Glenn

Management of fruit tree canopy temperature requires an accurate measurement of temperature, knowledge of whether temperature is limiting, and an understanding of temperature-modifying strategies and their economic and physiological consequences. The maximum photosynthesis of deciduous tree fruit crops occurs between 20 and 30°C when light is not limiting. Leaf temperatures above 30°C reduce photosynthesis; therefore, managing canopy temperature can increase productivity. The temperature of a tree's canopy is generally cooler than the air because transpired water evaporates and cools the canopy below air temperature. If water becomes limiting to the leaves, transpiration is reduced and leaf temperature rises because the leaf tissue is still absorbing radiation but evaporation of water is restricted.

Historically, the question of whether leaves are cooler than the air has been debated since the late 1800s. Technology limited a clear answer in the debate because leaf temperature could be measured only with mercury thermometers or thermocouples and only on single leaves. It was impossible to measure the temperature of an entire canopy until infrared (IR) sensors were developed in the 1960s. With the advent of IR thermometry, new theories developed and experimental evidence was collected to demonstrate that the surface temperature of a plant canopy could be predicted based on the micrometeorological conditions around the leaf and the effectiveness of the stomates to regulate transpiration rate. This body of work predicted that in humid environments, leaf temperature will equal or exceed air temperature, but in arid environments, leaf temperature can be as much as 7°C below air temperature. Understanding the mechanisms regulating leaf temperature led to useful techniques of quantifying plant stress, scheduling irrigation to meet the environmental demand on plants, and identifying new production practices that optimize canopy temperature and improve yield.

TEMPERATURE ASSESSMENT

Knowledge of the expected canopy temperature in a given environment draws attention to orchard management when the canopy temperature is warmer than predicted. Canopy temperatures warmer than predicted can be due to insufficient available soil water and are remedied by irrigation. Insect and disease damage can also interfere with leaf physiology, stomatal function, and the resulting transpiration rate that increases leaf temperature. A crop water stress index (CWSI) was developed as a measure of the relative transpiration rate occurring from a plant at the time of measurement, using plant canopy temperature (measured with an IR thermometer) and the vapor pressure deficit (a measurement of the dryness of the air). Jackson et al. (1981) present the theory behind the energy balance, and it separates net radiation from the sun into sensible heat that warms the air and latent heat that evaporates water from the leaf, e.g., transpiration. When a plant is transpiring without limitations, the leaf temperature is at a theoretically low baseline limit (generally 1 to 7°C below air temperature), and the CWSI is 0. As the transpiration decreases due to insufficient water or other adverse factors, the leaf temperature rises and can be 4 to 6°C above air temperature. As the crop undergoes water stress, the stomata close, and transpiration decreases and leaf temperature increases. The CWSI is 1 when the plant is no longer transpiring and leaf temperature reaches an upper baseline limit. In general, the crop does not need to be watered until the CWSI reaches 0.1 to 0.2. In this range, the crop is transpiring at less than the optimal rate, and plant performance will start to decrease. The use of the CWSI in tree fruit crops is not as direct as in agronomic crops, however, because the tree canopy often has gaps exposing wood, soil, and sky. The canopy must be uniform to establish accurate baselines and reliable field measurements for irrigation scheduling.

COOLING STRATEGIES

When canopy temperatures exceed 30°C, additional production practices are required to maintain high photosynthetic rates. Two general strategies are utilized to reduce the heat load on the tree: (1) evaporative cooling with water and (2) application of reflective materials. Evaporative cooling utilizes the latent heat of vaporization of water evaporation to cool the wetted leaf surface up to 14°C. Scheduling of crop cooling is based on air temperature and relative humidity and requires an automatically programmed irrigation system. Application rates vary from 3 to 10 millimeters of water per hour. Evaporative cooling maintains a thin film of water on the exposed canopy that evaporates and cools the leaves. Crop cooling and the subsequent reduction of heat stress increase yield, color development, and internal fruit quality. Evaporative cooling requires additional capital investment for the water distribution system, utilizes tremendous volumes of water, and necessitates that additional disease control be anticipated. When canopy temperatures do not significantly exceed 30°C, materials can be applied that reflect incoming solar radiation, thus lowering the heat load on the canopy. Kaolin is a common reflectant used in orchards. Some photosynthetically active radiation (PAR) is also reflected by the kaolin, but the value of reducing leaf temperature appears to overcome the loss of PAR on a whole-canopy basis, with similar fruit benefits as evaporative cooling. Reflectants have wide appeal because they can be applied with orchard sprayers at or prior to excessive heat periods and so do not require the capital and pumping costs of irrigation. Also, water resources are conserved. A benefit of both evaporative cooling and reflectants is the reduction of fruit sunburn damage.

Evaporative cooling can also be used in the dormant season to delay fruit bud development. Less water is required for evaporative cooling in the dormant season than in the growing season due to the reduced energy load of the environment. Overhead sprinkler irrigation, wind machines, and heaters are used in the spring to protect fruit flowers from freezing temperatures and are discussed in the chapter on spring frost control.

Tree canopy temperature management strategies utilize the physical properties of water and mineral materials to modify the microclimate of a fruit tree to reduce environmental damage. These strategies improve fruit quality and the stability of orchard productivity by reducing water stress, sunburn, and other temperature-related injuries.

Related Topics: FRUIT COLOR DEVELOPMENT; FRUIT GROWTH PATTERNS; IRRIGATION; LIGHT INTERCEPTION AND PHOTOSYNTHESIS; SPRING FROST CONTROL; TEMPERATURE RELATIONS; WATER RELATIONS

SELECTED BIBLIOGRAPHY

Faust, M. (1989). *Physiology of temperate zone fruit trees.* New York: John Wiley and Sons.

Glenn, D. M., G. J. Puterka, S. R. Drake, T. R. Unruh, A. L. Knight, P. Beherle, E. Prado, and T. Baugher (2001). Particle film application influences apple leaf physiology, fruit yield, and fruit quality. *J. Amer. Soc. Hort. Sci.* 126:175-181.

Jackson, R. D. (1982). Canopy temperature and crop water stress. *Advances in Irrigation* 1:43-85.

Jackson, R. D., S. B. Idso, R. J. Reginato, and P. J. Printer (1981). Canopy temperature as a crop water stress indicator. *Water Res.* 17:1133-1138.

Jones, H. G. (1992). *Plants and microclimate,* Second edition. Cambridge, UK: Cambridge Univ. Press.

Williams, K. M. and T. W. Ley, eds. (1994). *Tree fruit irrigation: A comprehensive manual of deciduous tree fruit irrigation needs.* Yakima, WA: Good Fruit Grower.

1. WATER RELATIONS
2. WILDLIFE

Water Relations

D. Michael Glenn

Water constitutes 80 to 90 percent of leaf and fruit tissue and more than 50 percent of an entire fruit tree by weight. Over 90 percent of the water entering a tree is lost through the leaves, and 95 percent of this water passes through the stomates, or pores in the leaves, which occupy less than 1 percent of the leaf area. The process of leaf water loss is termed transpiration, and a large fruit tree can transpire more than 400 liters of water in a day if the environmental demand for water is great. Over 90 percent of living cells are water because water is used for both chemical and physical requirements. Water is a medium for chemical reactions in the cell, is used in the chemical reactions of photosynthesis, and is an agent for the transport of chemicals in the diffusion process. Water is the medium for the transfer of carbon dioxide from the atmosphere into the mesophyll cells where photosynthesis occurs and for the transport of sugars from leaves to storage organs, such as fruit. Water is needed structurally for cell turgor and physical cell expansion. In addition, transpiration of water cools the leaves through the process of water evaporation.

PATHWAY OF WATER FROM THE SOIL TO THE LEAF

Water flows from the soil into the plant and out through the leaves in a continuous process that is controlled by the water demand of the environment surrounding the tree canopy (Figure W1.1). Water is absorbed through the root hairs, nonwoody roots, and to some extent by the woody root system. Water moves into the root through spaces in the cell wall and pores between cells, called the cortex, until it reaches the endodermis. The endodermis contains a suberized layer of cells,

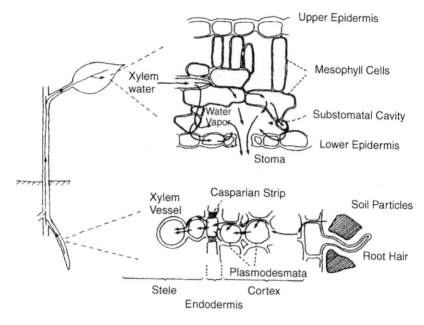

FIGURE W1.1. Movement of water from the soil to the leaf (*Source:* Modified from Jones, 1992.)

the Casparian strip, that blocks water movement unless the water moves through the cell membrane into the cell. Once water is in the cell, it can move through the Casparian strip through plasmodesmata into the stele, and in the stele, water moves back across a cell membrane into the xylem. The xylem is composed of vessel elements and tracheids that are functional only when their cells have matured and are dead. After the xylem cells are produced, the protoplasts are absorbed by adjacent cells. Before a cell dies, it builds a secondary cell wall that adds strength and prevents it from collapsing when tension develops in the xylem. Tracheids are longer and more narrow than vessel elements. Both cell types have pits in the sides or ends that allow water to flow from one cell to the next. In this manner, a continuous column of water is supported within the xylem from the stele, up the trunk of the tree, out the branches, and to the leaves.

ASCENT OF WATER IN PLANTS

Plants have evolved an elegant and highly effective means of transporting water from the soil to tremendous heights without the use of metabolic energy. The key is the surface tension of water. Water has a much higher surface tension than most other liquids because of the higher internal cohesion related to hydrogen bonding between the water molecules. If a column 10 micrometers in diameter and 3,000 meters tall were filled with water, it would hold the column of water and not drain as a result of the surface tension of water. Therefore, the transport of water is not limited by the height of fruit trees. The movement of water within a tree requires a continuous column of water; however, breaks in the column of water moving through the xylem do occur. Water stress occurs when the environmental demand for water exceeds the plant's ability to transport water to the leaves. As the demand for water begins to exceed the transport capacity, greater tensions develop within the xylem, and at high tensions, columns of water in tracheids and vessels break, leaving cavities. A bubble of air is generally contained within the individual element. Some cavitations in the xylem permanently block water movement through that xylem element. In other xylem vessels and tracheids, the cavitation may be filled when the tree is rewatered through irrigation, or it may fill overnight due to root pressure. Root pressure develops in plants because water is drawn through the endodermis by the concentration of salts and organic molecules dissolved in the stele. Water moves into the stele by osmosis, in which the endodermal cells are the selectively permeable membrane. The water that moves into the stele is forced up the xylem, and it is important for refilling the xylem during the night when transpiration ceases.

TRANSPIRATION STREAM OF WATER

The transpiration stream of water literally "pulls" water out of the soil as water moves from the xylem elements into the tree canopy—driven by the energy available to evaporate water. As water flows, it encounters resistances from the soil, moving through the root cortex and membranes, passing along and through the xylem vessels, changing phases to a gas in the stomatal cavity, and passing through the stomata into the air. The flow of water from the soil to the root surface is generally not limiting until either the soil dries and shrinks, causing gaps between the root system and the soil, or the water films around soil particles become so thin that hydraulic conductivity to the root is reduced. Water movement through the external root cortex is not limiting, but when water moves through the endodermal cell membrane to pass through the Casparian strip and then moves through a second membrane to transfer into the xylem, there is considerable resistance due to the cell membrane. Water flow in the xylem is generally not limiting unless extensive cavitation or blockage has occurred due to water stress or biotic stress such as disease or insect damage. The stomatal cavity and opening are the final resistance in the flow of water, and the plant can control the size of the stomatal opening and, hence, the rate of water vapor transport from the leaf. The size of the stomatal opening, or stomatal conductance, is related to environmental and biotic factors. Stomatal aperture generally reaches a maximum at 25 to 50 percent of full sunlight, and the tree will maintain high conductance unless other factors cause stomatal closure. Stomates tend to open as the leaf temperature rises and close as the relative humidity decreases. Water stress will cause stomates to close. Water stress develops when the energy to evaporate water exceeds the transport of water to be evaporated. The resulting deficit of water results in tension developing in the xylem column of water. Stomatal closure reduces the transport of water from the leaf and allows time for water transport from the soil to reduce the tension. When the tension exceeds a threshold, the column of water begins to break within the xylem vessels. The cavitation of the xylem vessels further reduces the transport rate and supply of water to the canopy and results in plant wilting, stomatal closure, and the reduction of photosynthesis.

COMPONENTS OF PLANT WATER RELATIONS

The flow of water from the soil to the leaves not only cools the canopy through transpiration but also supplies water to all living cells in the plant. Cells have direct contact with films of water from the xylem. The water films have a measurable "tension" that is called the water potential (Ψ_w), and this tension is developed when the transport of water does not meet the environmental demand for evaporation. Cells are bathed in water at a Ψ_w that ranges from approximately zero under well-watered conditions and no water stress to a negative pressure (less than zero) when water stress is developing. The cell, however, must maintain a positive pressure, or turgor, for expansion and normal biochemical function. Cell turgor pressure (Ψ_p) is the result of three primary factors:

1. *Solute concentration, or solute potential* (Ψ_s): The solutes in the cell draw water into the cell via osmosis. The Ψ_s of the cell is less than zero, and the more negative its value, the greater is the potential influx of water.
2. *Effect of water-binding colloids and capillary attraction for water, or matric potential* (Ψ_m): Water is held by electrostatic forces to charged surfaces in the cell, such as proteins and nucleic acids, and the capillary channels within the cell wall also bind water. The Ψ_m in the cell is generally of little significance in maintaining turgor because the volume of water related to it is very small.
3. *Effect of gravity* (Ψ_g): This effect is generally negligible except when comparing water potentials at different heights in a tree.

Cell water relations can be expressed algebraically as:

$$\Psi_w = \Psi_s + \Psi_m + \Psi_p + \Psi_g$$

Assuming that matric and gravitational potentials are negligible and cell volume does not change, the following illustrates the relationships between the components of cell water:

Condition of the cell	Ψ_w	=	Ψ_s	+	Ψ_p	Units
No water stress and the cells are fully turgid	0	=	−2	+	2	MPa
Moderate water stress and the cells are partly turgid	−1	=	−2	+	1	MPa
Severe water stress and the cells are flaccid	−2	=	−2	+	0	MPa
Severe water stress and the cells are partly turgid	−2	=	−3	+	1	MPa

In this example, if the concentration of solutes in the cytoplasm (Ψ_s) increases, then turgor pressure will increase, at any constant negative Ψ_w . This is a common adaptation in plants called osmoregulation that is one of many ways plants adapt to their environment. Water stress and plant water relations are very complex phenomena, and this brief explanation only highlights some general trends on a whole-plant basis.

Yield and plant growth are reduced more by water deficits than by any other limiting factor in a plant's environment. Daily water deficits that occur during hot periods of the day, as well as seasonal deficits of water, alter a plant's morphology, physiology, productivity, and quality as a food product. Water deficits also increase a plant's susceptibility to insect and disease damage. An understanding of plant water relations aids in diagnosing conditions that limit plant growth and development. Plant breeders utilize knowledge of how plants morphologically, biochemically, and physiologically adjust to water stress in order to adapt new cultivars to their environment.

Related Topics: CARBOHYDRATE PARTITIONING AND PLANT GROWTH; IRRIGATION; TEMPERATURE RELATIONS

SELECTED BIBLIOGRAPHY

Faust, M. (1989). *Physiology of temperate zone fruit trees.* New York: John Wiley and Sons.
Jones, H. G. (1992). *Plants and microclimate,* Second edition. Cambridge, UK: Cambridge Univ. Press.
Kramer, P. J. and J. S. Boyer (1995). *Water relations of plants and soils.* New York: Academic Press.
Nobel, P. S. (1991). *Physiochemical and environmental plant physiology.* New York: Academic Press.

– 2 –

Wildlife

Tara Auxt Baugher

Wild animal diversity is an important component of a healthy orchard ecosystem. However, if overabundant, certain species may cause considerable economic loss. According to the U.S. Department of Agriculture Animal and Plant Health Inspection Service, estimated wildlife damage to U.S. agriculture exceeds $550 million annually. Common wildlife problems in tree fruit plantings are fruit-eating birds, such as starlings *(Sturnnus vulgaris),* and browsing mammals, such as voles *(Microtus* species), deer *(Odocoileus* species), and rabbits *(Sylvilagus* species). Strategies to maintain a balance between human and wildlife needs vary from one agroecosystem to the next.

INTEGRATED APPROACH TO WILDLIFE MANAGEMENT

Integrated pest management (IPM) is a broad-spectrum approach to limiting wildlife damage that has been adopted by many fruit growers. Several methods of control are integrated simultaneously or alternately, and the key to success is routine monitoring. Wildlife control procedures are integrated with practices to manage all classes of pests, including insects, diseases, and weeds. For example, research by I. A. Merwin at Cornell University demonstrates that hay/straw or fabric mulch increase the potential for vole damage and therefore should not be used for weed control in situations where voles are a threat. Effective monitoring entails assessing wildlife and predator populations, wildlife habitats and behaviors, damage patterns, possible impacts on nontarget organisms, and various conditions that may influence control efficacy. An example of a widely

367

used monitoring tool is the apple sign test, developed by R. E. Byers of Virginia Polytechnic Institute, for estimating the potential for meadow or pine vole damage in an orchard block. With timely monitoring, management tactics can be employed prior to the establishment of animal feeding or browsing habits.

GENERAL CONTROL STRATEGIES

Wildlife management strategies fall under six general categories:

1. *Natural control* is encouraged by providing adequate nesting, denning, and perching sites for predators.
2. *Habitat modification* includes various changes in orchard culture to discourage wildlife damage.
3. *Exclusion* is the use of fencing or protective barriers and is generally the most reliable but also the most costly control tactic.
4. *Repellents* are taste- or odor-based materials that inhibit damaging behavior and are cost-effective in orchards with low damage potential.
5. *Scare tactics* include visual scare devices and noisemakers to discourage bird feeding and the use of guard dogs to deter deer.
6. *Population reduction* options are trapping, baiting, or hunting and generally should be discussed with a wildlife conservation officer.

Optimum success results from combining an array of strategically timed management plans. If a vertebrate pest population is small, damage potential can sometimes be maintained at a low level with a program combining habitat modification and natural control. In situations where pest populations are already high, these two strategies must be integrated with additional control measures.

Fruit growers recognize wildlife as a valuable resource, and many set aside animal refuge areas. On the other hand, the consequences can be devastating if a wildlife damage monitoring and control program is not established prior to planting a new orchard. Budgets for preventing wildlife damage should be based on estimated impacts on profitability over the life of an orchard rather than a single year. Animal damage to agricultural crops is a complex issue, and the ultimate goal is to develop an ecological framework for wildlife stewardship.

Related Topics: PLANT-PEST RELATIONSHIPS AND THE ORCHARD ECOSYSTEM; SUSTAINABLE ORCHARDING

SELECTED BIBLIOGRAPHY

Baugher, T. A. (1986). Deer damage control: An integrated approach. *Compact Fruit Tree* 19:97-102.

Byers, R. E. (1984). Control and management of vertebrate pests in deciduous orchards of the eastern United States. *Hort. Rev.* 6:253-285.

U.S. Department of Agriculture Animal and Plant Health Inspection Service (1997). *Managing wildlife damage: The mission of APHIS' wildlife services program,* Misc. pub.1543. Washington, DC: USDA.

Index

Page numbers followed by the letter "f" indicate a figure; those followed by the letter "t" indicate a table.

T - #0358 - 101024 - C6 - 212/152/23 - PB - 9781560229414 - Gloss Lamination